AF371679

Reproduced on the cover is an electron micrograph of a sympathetic ganglion and plexus showing nerve cell bodies and their processes. (x 4,500). Courtesy Michael Ross, Department of Cell Biology, School of Medicine, New York University.

George Camougis is President and Research Director of New England Research and Development Laboratories, Worcester, Massachusetts, and Professor Affiliate of Physiology at Clark University there. He holds a B.A. degree from Tufts University and M.A. and Ph.D. degrees from Harvard.

NERVES, MUSCLES,
AND ELECTRICITY

NERVES, MUSCLES, AND ELECTRICITY:

An Introductory
Manual of Electrophysiology

George Camougis

New England Research and Development Laboratories
and
Clark University

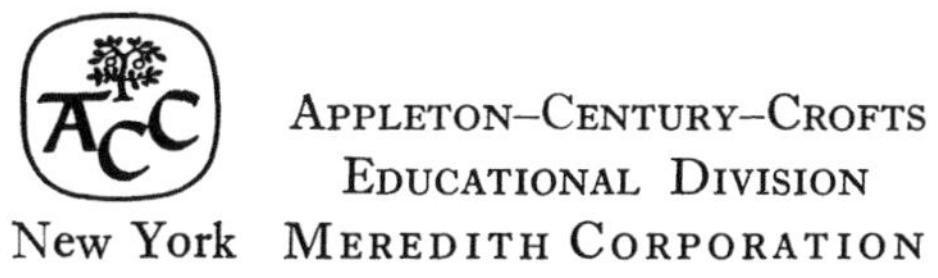

APPLETON–CENTURY–CROFTS
EDUCATIONAL DIVISION
New York   MEREDITH CORPORATION

ISBN 978-1-4684-1371-7   ISBN 978-1-4684-1369-4   (eBook)
DOI 10.1007/978-1-4684-1369-4

Library of Congress Card Number: 79-117053

390-16670-7

PREFACE

For centuries man knew about the lightning of the sky (atmospheric or physical electricity) and the numbing effects from contact with powerful electric fishes (animal electricity). Then, after proper experimentation and a synthetic rationale, it began to appear that physical and animal electricity were related in fundamental respects. This realization was made at the end of the eighteenth century, since when the pages of history have been replete with exciting discoveries and developments in electricity and magnetism, electrochemistry, and electrophysiology. It is hoped that this manual will enable some students to relive some of that excitement. The author remembers vividly the excitement when, as an undergraduate, he saw his first action potential.

This book is not intended for any particular group of students; it should prove to be of some value to students in secondary schools, colleges, and graduate schools. Based on personal experience, the author feels that many teachers will also find it of use. Nor is the manual intended solely as a laboratory manual for an introductory course in neurophysiology. Some of the experiments might be introduced into the formal laboratory schedule of general or animal physiology courses. Alternatively, the various experiments might provide the bases for the beginning of special projects lasting for a full semester or even longer. The author is well aware of the plethora of works in technical aspects of electrophysiology, but he is equally aware of the subsequent frustration of some beginners directed to read them. Perhaps in this simpler manual, then, such beginners, irrespective of academic circumstances, may find a friend.

The experiments have been selected for simplicity and reliability, and include small laboratory animals which require minimal expense and facilities. The complexity of the instrumentation progresses with the experiments. An extensive bibliography is included to assist the reader in advancing to more involved techniques. Finally, discussions on instrumentation, interpretation of results, and the selection of apparatus are included to widen the practical scope of the manual. It is hoped that some readers will be stimulated to progress beyond its limited pages, for only then will the manual really have justified its existence.

The author acknowledges with pleasure the contributions of a myriad of teachers, students, and colleagues in various laboratories over the years. Without their help and enthusiasm these pages would be impossible. The author also expresses deep appreciation for the able assistance of Miss Winifred J. Mulhern in the preparation of the manuscript.

G. C.

Worcester, Massachusetts

CONTENTS

Part One

INTRODUCTION

Electrophysiology is not a basic science in the same sense as the established fields of biology, chemistry, and physics. It is not even a well defined subdiscipline. It has no body of theory peculiar to itself; rather it is an assemblage of techniques and skills, and draws heavily from biology, chemistry, physics, and engineering for its principles.

What, then, is electrophysiology? Various definitions are to be found in the literature. For our purposes, let us adopt the following definition: electrophysiology concerns itself with the study of (1) electrical phenomena in living systems, and (2) what happens when electricity is applied to living matter. Very often electrophysiology is used instead of the word "neurophysiology." Neurophysiology includes in its arsenal more than electrical apparatus for its attacks on the problems of studying how nerves function. However, because electrophysiology techniques were used in the very early studies of nerve and muscle, the terms "electrophysiology" and "neurophysiology" have come to be used interchangeably. It must be emphasized that this is a manual of electrophysiology, concerning itself only with limited aspects of electrophysiology: its main concern being the detection and display of certain bioelectrical phenomena, such as are chiefly to be found in nerve and muscle tissue, but also are found in many organisms, including plants. Indeed, one can question whether the dynamism of life is possible without electrical correlates.

Bioelectrical phenomena can be classed into two convenient divisions: steady or *tonic* potentials, and alternating or *phasic* potentials. Potential is used

1

to mean electrical potential, voltage, or electromotive force (e.m.f.). Tonic potentials are potentials which either do not vary in time, or vary so little that they are virtually steady. Perhaps the best known example of a steady potential is the potential difference across the membrane of a resting cell. Consider, for example, a muscle cell as shown in Fig. 1. This microscopic cell has a potential difference of about 70 mV (70/1000 volts) across the cell membrane; this is known as the *resting potential* or *transmembrane* (also *membrane*) *potential*. As shown in the figure, the resting potential can be measured by impaling a cell with an appropriate microelectrode and connecting it to a recording instrument whose other terminal is connected to an external reference electrode. The outside of the cell is positive.

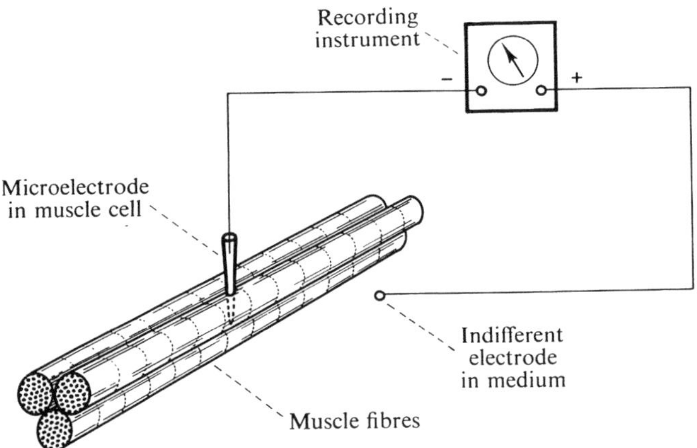

FIG. 1. Diagram of a muscle fiber impaled by a microelectrode to measure the membrane potential. The indifferent electrode is in the external saline medium which bathes the cells.

Another type of steady potential exists following the injury of a cell or tissue. Let us consider the muscle again. The muscle is a tissue, and is made up of many parallel muscle fibers or cells. A cut across the belly of the muscle (Fig. 2) disrupts the integrity of the cells. This in turn disrupts the transmembrane potentials, and, therefore, the vicinity of the cut is characterized by a *depolarized* condition. The healthy part will then show a positive potential with respect to the cut or injured part. This potential is called an *injury potential* or *demarcation potential*. Since injury potentials involve tissues rather than cells, and hence simpler manipulative and measuring techniques, it is easy to understand that, historically, they were detected and measured before membrane potentials.

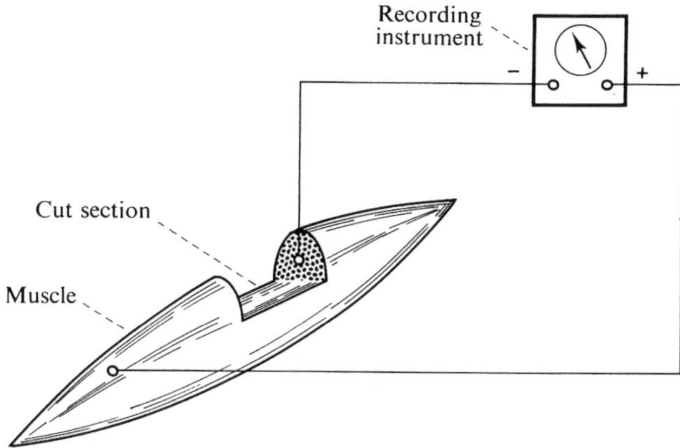

FIG. 2. Diagram of an experimental method to show a demarcation potential in muscle. The cut area is electrically negative relative to the intact area.

In addition there are steady or d.c. potentials between different parts of the body of animals. For example, the salamander has various d.c. potentials of up to 30 mV along the spinal axis and the extremities, these sites being negative with respect to a point on the head (Becker, 1960). Such potentials have not been studied as intensively as membrane potentials or action potentials, which are discussed next.

A phasic potential is a varying phenomenon; it represents a relatively rapid fluctuation of potential from a functioning cell, tissue, or organ. Examples would include electrical potentials from muscle (*electromyogram* or EMG), nerve (*electroneurogram* or ENG), the heart (*electrocardiogram* or EKG), and the retina (*electroretinogram* or ERG). The essential point is that the electrical potential of a certain location varies in time relative to another location. The *action potential* from peripheral nerve is a good example of a phasic potential. As is shown diagrammatically in Fig. 3, action potentials are *self-propagating* along the long axes of nerve and muscle fibers. Note that as the action potential (the *depolarized* region) moves relative to the recording electrodes, the electrodes "feel" corresponding differences in potential relative to each other. Sometimes in electrophysiology we have one electrode on the active site (*active electrode*) while the other electrode (*indifferent electrode*) is placed at a more remote site. This is discussed further in Part Three. Additional information on electrical phenomena in nerve and muscle may be found in appropriate references in Table 1.

As with many scientific and technical specialties, methodology is the heart of electrophysiology. Progress in electrophysiological research has been possible

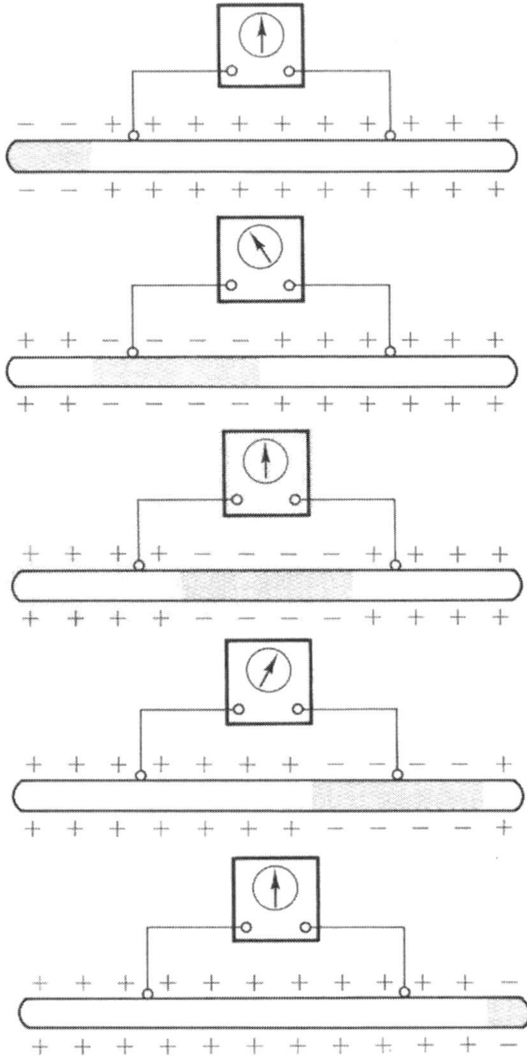

FIG. 3. Conceptualized diagram of a propagated action potential (stippled area) moving along the long axis of an excitable tissue from left to right. The area of negativity registers first at one electrode and later at the other, thus causing biphasic movement of the recording instrument.

TABLE 1. A SURVEY OF GENERAL REFERENCES IN NEUROPHYSIOLOGY, ELECTROPHYSIOLOGY, AND BIOMEDICAL INSTRUMENTATION

Topics	References	
General neurophysiology and electrophysiology	Tasaki, 1953; 1968 Brazier, 1960 Ruch *et al.*, 1961 Suckling, 1961 Galambos, 1962 McLennan, 1963 Roeder, 1963	Wooldridge, 1963 Eccles, 1964 Hodgkin, 1964 Ochs, 1965 Florey, 1966 Katz, 1966 Stevens, 1966
Electrophysiological methods and experiments	Dickinson, 1950 Donaldson, 1958 Welsh & Smith, 1960 Burns, 1961 Nastuk, 1963 Nastuk, 1964	Goldstein, 1964 Whitfield, 1964 Smith, 1966 Bures *et al.*, 1967 Hoff & Geddes, 1967 Suckling & Koizumi, 1967
Biomedical electronics and instrumentation	Whitfield, 1953 Catton, 1957 Brown & Saucer, 1958 Stacy, 1960 Rosenblith, 1962 Fogel, 1963 Brown & Webb, 1964 Camishion, 1964 Kay, 1964	Newman, 1964 Tolles, 1964a Hill, 1965 Yanof, 1965 Phillips, 1966 Rushmer, 1966 Suprynowicz, 1966 Offner, 1967 Geddes & Baker, 1968

through rapid developments in methodology and instrumentation. A beginner who visits an electrophysiology laboratory is usually baffled by the array of instruments, wires, dials, switches, meters, etc. Oftentimes, to add to the confusion, idle instruments and apparatus may be intermingled with the units in actual use. The second chapter of this manual is devoted to discussing the individual components commonly used in electrophysiological research, the third to examining how they fit together into instrumentation systems, and part three to some simple experiments in electrophysiology. Since it is impossible to separate techniques from other aspects of electrophysiology, the only way to learn is by doing experiments. As it becomes necessary for the reader to get involved in more advanced methodology (e.g., cortical evoked potentials; signal processing), additional reading will be required. Table 1 includes an extensive survey of references in electrophysiological methods and in biomedical instrumentation. To summarize, electrophysiology is the study of electrical phenomena in living tissues; instrumentation is what makes such study possible.

Part Two

INSTRUMENTATION

*I often say that when you can measure what you
are speaking about, and express it in numbers,
you know something about it.*—LORD KELVIN

Chapter One

Basic Principles

We may start off our discussion of instrumentation by recalling two facts. First, in electrophysiological techniques the standard laboratory approach is to detect and measure bioelectrical potentials, and second, by far the majority of bioelectrical potentials are so small that sensitive apparatus is required to detect and measure them. In addition, we may add that electrophysiology includes the investigation of how living tissues react to electrical currents. We have now standardized some of our knowledge on the applications of such currents; consequently electrical stimulators are available for investigations with electrically active tissues. Finally, we should remember that our living systems are essentially watery saline systems that, through a variety of electrodes, are coupled to physical instruments. Briefly, then, the components of our instrumentation will include *electrodes, amplifiers, read-out devices,* and *stimulators.* These components are discussed in the next section.

There are numerous biological phenomena which are nonelectrical in nature; indeed, most of the phenomena associated with vital processes involve little or no electrical correlates of sufficient magnitudes to be useful in measuring. Yet the most refined instrumentation techniques depend upon the resolution of a given phenomenon into an electrical magnitude which can be fed into the instrumentation system. There are various devices called *transducers* which can convert nonelectrical phenomena into electrical phenomena. These are very useful in biological research; they are also discussed in the next section which covers the various components, but before proceeding, let us consider briefly the role of instrumentation in general. The familiar quotation above from Lord Kelvin emphasizes the essential point; one wishes to measure so that one really begins to understand the particular phenomenon in question. This concept is fundamental to all science and technology. The growth of modern instrumentation has come about from the need to gather accurate quantitative data. In the

physical sciences we find that instruments allow us to do at least three things:
  (1) detect phenomena
  (2) measure phenomena
  (3) control phenomena

A logical extension of these operations leads us to the components of a typical instrumentation system used in electrophysiological research. Thus we see units for transduction and/or detection, amplification, display and/or recording, and stimulation. The investigator must remember that in the science of instrumentation the laws governing the operation of an instrument and the specifications and response characteristics of an instrument must be known. Only then can the relationship between input and output signals be understood fully. The specifications of a given instrument must be studied carefully before purchase, in order to establish its function and ability, and then, after acquisition, the specifications must be studied again, this time in conjunction with the operating instructions. One would be amazed to know how these principles of common sense are neglected. In the sections that follow, certain specifications of interest to electrophysiology research will be discussed. Unfortunately, many commercially available instruments are over-sophisticated for routine experiments in electrophysiology. This often acts as a deterrent to teachers and investigators who might otherwise include electrophysiological procedures in their activities. See Appendix 2 for comments on the selection of apparatus.

Chapter Two

Typical Components

Electrodes. "Electrode" is a general term for a device that couples the biological preparation to an electrical instrument. Electrodes are simple components in comparison with the other units required; however, they can offer numerous problems in technique, so it is impossible to overstress the attention that must be given to them.

Electrodes used for biological work may be classed into *stimulating* electrodes, and *recording* electrodes. Stimulating electrodes usually consist of silver wires in contact with the biological preparations. Since most stimuli are brief, square-wave pulses, no problems of polarization occur. Details about the gauge, length, and configuration of the electrodes depend upon the experiment being done. Common sense as much as experience will dictate the final design. Electrodes will rarely be over a few inches in length, therefore, they are connected to the stimulator by longer leads to which they are coupled, usually by plugs or clips.

Similar general remarks apply to recording electrodes, whose precise design depends greatly upon the preparation, types of potentials, and the localization requirements. Again, since most potentials are small and phasic in nature, polarization effects will pose no problems. Plain silver wire is the most common type of recording electrode, which needs to be adequately strong and yet flexible enough to be manipulated, because in some experiments it is convenient to fashion small hooks at the ends of the electrodes (Fig. 4). Silver wire from about 20 to 30 gauge will be adequate for most stimulating and recording needs.

Whenever d.c. passes through a metallic wire in contact with the biological preparation, polarization of the electrodes takes place. This happens because the oppositely-charged ions of the biological fluids are attracted to the respective electrode poles. The ion accumulation at the electrodes affects the current and

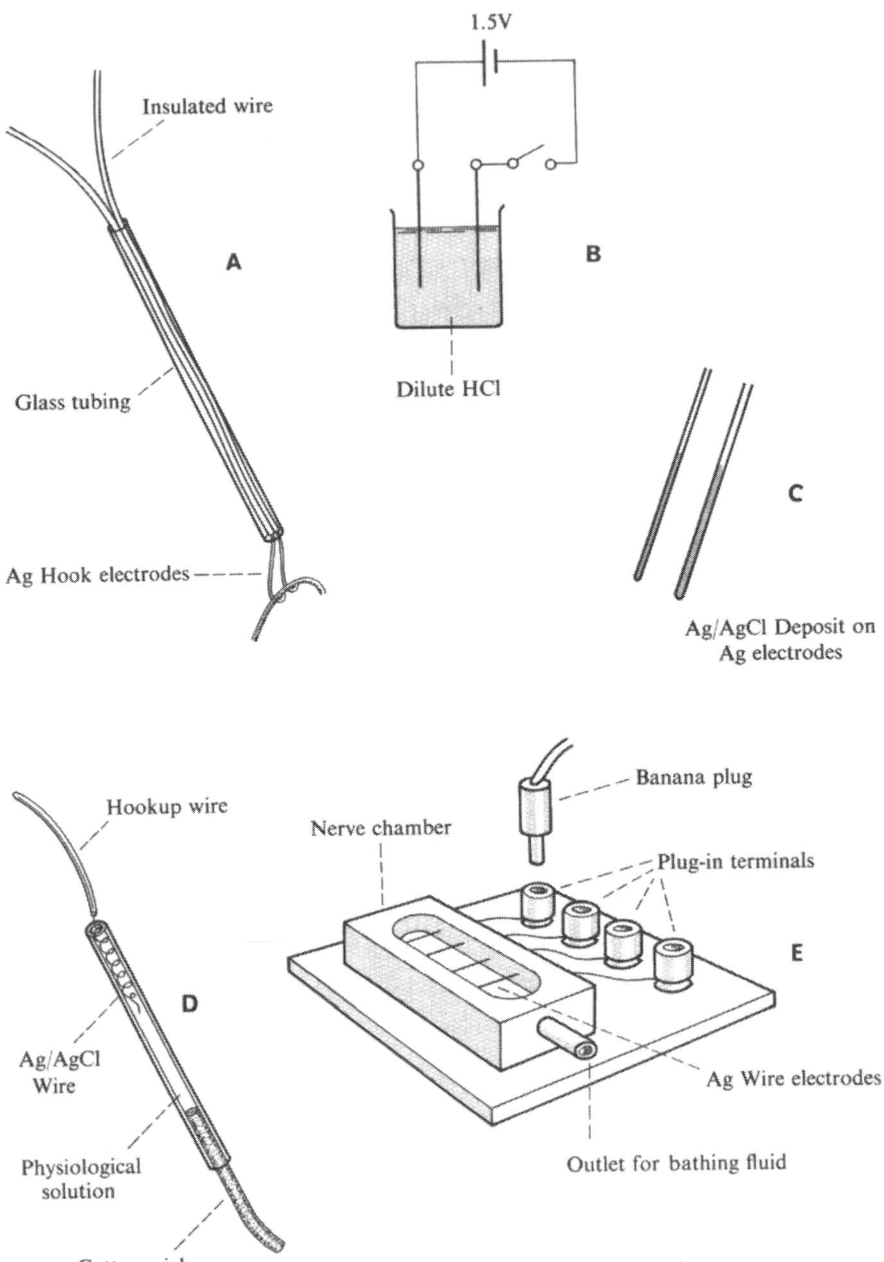

FIG. 4. Semi-diagrammatic illustrations of various electrodes and accessory components.

A. Simple silver wire hook electrodes connected to insulated hookup wire, and placed inside glass tubing for ease of manipulation. B. Diagram of arrangement to make Ag/AgCl electrodes. C. Ag/AgCl electrodes. D. Cotton wick type of non-polarizable electrode. E. Nerve chamber with silver wire electrodes for mounting the nerve preparation.

voltage values of the circuit, thus leading to errors. For that reason, nonpolarizable (or nonpolarizing) electrodes are often used in electrophysiology whenever appreciable d.c. components are anticipated. Two types of nonpolarizable electrodes are recommended because of simplicity. One is the Ag/AgCl type, made by placing the electrode in dilute HCl and passing a weak d.c. through it. The current is applied briefly, until a definite gray-brown deposite of AgCl film is formed on the silver electrode (Fig. 4): such an electrode is non-polarizable. The other type is the wick electrode. A piece of cotton wick or a thick thread makes contact with the preparation at its tip while the other end is in contact with a physiological solution inside a tube (Fig. 4). An Ag/AgCl wire serves as a connection from the solution to the lead going to the particular instrument. If greater mechanical rigidity than that of the wick is required, a piece of balsa wood may be used. Quickly the balsa becomes saturated with solution and acts as a suitable electrode.

Even after the appropriate wires for electrodes are available, often the need arises for an electrode assembly in some sort of chamber—especially true with isolated tissue work (e.g., Experiments 7 and 10). Many different electrode assemblies attached to nerve chambers have been described. A simple example is shown in Fig. 4. Such chambers should be simple and appropriate for a given experiment. More highly specialized chambers may be designed at a later date as the experimentation gets more involved.

Two additional points will be stressed about electrodes. First of all, good electrical contact must exist between electrode and tissue. Many problems in electrophysiological techniques would be solved if this simple rule were always observed. Second, excessive fluid between electrodes, especially recording electrodes, will shunt out the signals. The resistance path through the saline fluid is very low compared with the input of an amplifier, so that the signal is not picked up, thus giving the erroneous impression that the preparation is not active.

Many specialized electrodes and electrode assemblies have been designed over the years, various types of microelectrodes having been used extensively more recently. For clinical work, special electrodes for EEG and EKG recording are used. All such special applications are outside the scope of this elementary manual. For additional discussions on electrodes, the reader may consult Dickinson (1950), Donaldson (1958), Welsh and Smith (1960), Burns (1961), Goldstein (1964), Kay (1964), and Nastuk (1964).

Amplifiers. The function of an amplifier is to increase the voltage of a bioelectrical signal so that it can be displayed or further processed by another instrument (Fig. 8). If the reader does not know the fundamentals of amplifier electronics, he is urged to read up on the subject. The necessary information is available in basic textbooks on electronics, and additionally, several of the references in Table 1 have basic discussions on amplifier electronics and circuit design (e.g., Whitfield, 1953; Brown and Saucer, 1958; Donaldson, 1958;

Camishion, 1964; Yanof, 1965; Phillips, 1966). How far one should go in the study of electronics before doing electrophysiological research is a moot question: in the opinion of the author, it really is a question of personal philosophy. Meaningful studies in electrophysiology can be done without expertise in electronics; however, a basic knowledge of what each instrument can do is doubtless important for proper technique. The important thing to remember is that an instrument is a tool to do research; too much concern over the tool may result in neglect in doing the experiment.

Not only must an amplifier increase the amplitude of a signal, it must do so without distortion. Most bioelectrical amplifiers are high-gain amplifiers designed for minimal distortion. Faithful reproduction of the waveform being vital for proper investigation, good amplifiers are needed for electrophysiological investigations. Most amplifiers are designed to feed into read-out devices which themselves have amplifiers (e.g., cathode ray oscilloscope or graphic recorder), and are more properly termed preamplifiers because of their design. Let us consider briefly certain specifications of interest in biological preamplifiers.

Total *amplification* is an obvious point of interest. Most general purpose preamplifiers for biological applications will have amplification values in the range of 1000 to 10,000. When combined with the amplification of the read-out device, this will provide adequate gain for practically all applications. Generally speaking, full amplication of the preamplifier is not used unless needed; the amplifier of the read-out device is used to provide additional gain for good recording.

Another specification is the *frequency response* of the instrument. Most preamplifiers have linear responses from a few Hz to at least 10 kHz, and in addition, various *filters* are available which can be set at appropriate values. For example, in recording a nerve action potential, the lower frequency response might be set at 1 to 10 Hz and the high frequency response set in the range of 1 to 10 kHz, exact settings depending upon the manufacturer of the instrument. Such filter values will eliminate much low frequency and high frequency noise unrelated to the biological action potentials. If too much filtering is used, there is danger of distorting the signal. Such distortion may be a big problem when the shape of the waveform is desired. Otherwise, if one is interested only in the occurrence of a bioelectrical event (e.g., the presence of an EKG), concern with distortion is less important.

One of the values that the experimenter must determine is the amplitude of the potentials he is recording (see Chapter 7), so some sort of *calibration* for the recording system is needed. Many preamplifiers have reference calibration signals which can be substituted for the input signal, a typical calibration signal being 500 μV. In general, the calibration should be done with signals of the same order of amplitude as the bioelectrical signals being studied. If the read-out device has precise calibrations on the gain control, a reference source in the preamplifier is unimportant. For example, if an oscilloscope is set at

100 mV/cm, a signal of 50 mV, irrespective of its source, will cause a deflection of 5 mm on the screen. If required, calibrators for a reference source are available commercially; a simple calibration signal may be obtained, however, by using a 1.5 V battery and appropriate precision resistors arranged in a voltage divider network.

The nature of the input coupling is another specification of concern. Amplifiers are either *a.c.-coupled* or *d.c.-coupled,* these terms being derived from the characteristics imposed by circuit elements located between the input signal and the grid of the first input stage. A.c.-coupled instruments have capacitance input, and therefore d.c. or steady state potentials are rejected as inputs. Tonic potentials cannot be recorded with such instruments. On the other hand, d.c.-coupled instruments will accept both phasic and tonic signals. However, d.c.-coupled amplifiers are more apt to produce baseline drift, and are usually more expensive. Unless the recording needs require d.c. input, most research in electrophysiology is done with a.c.-coupled preamplifiers. The amplifiers of most oscilloscopes have both a.c. and d.c. coupling, but d.c. input is not necessary for most biological applications. This will be discussed further.

A final point of discussion also centers on input coupling. The two input leads to the amplifier will pick up any signal of differing potential. In doing electrophysiology research, the investigator has two conditions with which to contend. First of all the bioelectrical signals are very weak and high amplification must be used, and second, all sorts of stray electromagnetic signals are radiated by lights, instruments, remote radio stations, etc., to which the input side of the preamplifier is not immune. The dilemma is obvious; both types of signals are picked up and amplified, but somehow the unwanted signals or *noise* must be rejected. If one input lead is grounded, the other will pick up all potentials, including the noise, whose values differ relative to ground: such recording is known as *single-sided* recording. On the other hand, it is possible to design the amplifier so that signals coming in phase (such as 60 Hz or other noise) to the input leads will act on the grids to two input tubes and they will cancel each other. This is referred to as *push-pull* or *differential* input. With such recording, neither input electrode is at ground potential, and they are fed into the two differential input terminals. The noise is very likely to produce potentials of about the same order at both leads, while the bioelectrical potentials are localized in such a way that, at a given instant, the electrodes are at differing potentials. Thus, signals *in phase* (e.g., noise) are cancelled, while signals *out of phase* (e.g., action potentials) are added by the push-pull operation. However, push-pull amplifiers are not perfect, and a percentage of the noise signal passes through to the recording system, but this percentage may be small enough to be negligible. Differential input preamplifiers include a measure of this in their specifications; it is called *common mode rejection ratio.* Such ratios are usually several thousand to one. Thus, push-pull amplification combined with proper shielding (to be discussed later) will enhance the probability of excellent recording technique.

For those intending to do serious research in neurophysiology, further reading about amplification is important. Principles and practice of amplification applied to biological problems are discussed extensively by Whitfield (1953, 1964), Donaldson (1958), Suckling (1961), Kay (1964), and Nastuk (1964).

Read-Out Devices. So that an electrical output signal may be useful to us, it must be displayed in such a way that we know something about it. We cannot really visualize what an amplifier is doing until its output can be fed into an instrument that can display the signal or some parameter of the signal. Such a device is called a *read-out* device.

Before discussing read-out devices, let us consider once again the nature of our bioelectrical potentials. If the signal is a tonic potential it may be measured by any instrument which will accept d.c. and which has sufficient sensitivity. Phasic action potentials, on the other hand, may not be readily displayed if they are very fast; for instance, a nerve action potential lasts only a little over 1 msec. Since the upper frequency limit of response is set by the moment of inertia of the mechanical components, graphic recorders cannot respond quickly enough to record a nerve action potential. For that reason, most action potentials are displayed on a cathode ray oscilloscope (CRO). As the technology of graphic recording improves, higher frequencies of response are achieved; already, for example, an instrument is available that will respond in the kHz frequency range. However, the CRO is still the main read-out device for electrophysiological recording, and, since most of the experiments described in this manual require its use, the discussion emphasizes it over other read-out devices.

The heart of the CRO is the cathode ray tube (CRT). This consists of a system of electrodes and plates to change the path of a high velocity electron beam by electrostatic deflection. The reader unfamiliar with the CRT should read an elementary account of its operation and construction in a basic text in electronics; further accounts may also be found in Dickinson (1950), Whitfield (1953), Brown and Saucer (1958), Suckling (1961), and Camishion (1964). The mutually perpendicular plates can move the electron beam in two dimensions as displayed on the phosphor screen of the tube, thus giving a graphic tracing of the voltages on the plates. The CRO, then, can be looked upon as a CRT with associated circuitry to enable a useful display of a graph on the screen.

Let us now consider the two axes of this graph. For biological applications, the most useful arrangement is to have time as the horizontal axis. This is done by the sawtooth generator which imparts a given rate of displacement of the beam from left to right across the screen. At the end of its excursion, the beam is switched instantaneously back to the beginning of its time cycle and the process is repeated: this recurs rather quickly so that the frequency of this cycle may be many times per second. The other axis, the vertical axis, is a voltage axis. The magnitude and polarity of this voltage is determined by the vertical deflection caused by the input signal going to the amplifier associated with the plates. In summary, then, we have the horizontal axis (time) and the vertical axis (voltage) as the two essentials of our graphic display. The various knobs and dials of the CRO concern us only to the extent that they permit us to get a good trace and to control the axes.

One of the first specifications of interest is the *sweep circuit frequency* range. Most CRO's have upper frequencies far exceeding needs for biological applications (e.g., in the megacycle range). The lower frequencies may be too high (several cycles per second), and often a very slow trace is needed for some applications. However, most general purpose CRO's will have an adequate sweep frequency range for conventional experiments, the sweep speed being specified in cycles per second or in cm per unit time. The frequency is adjusted so that appropriate display for the particular potentials is achieved. The sweep speed may be relatively high (Experiment 10) or relatively slow (Experiment 6). Adjusting the proper sweep speed usually offers few problems.

The next item of interest is the *sensitivity*, usually given in V/cm or V/in, of the vertical amplifier. A high gain amplifier may have a sensitivity of greater than 1 mV/cm. With such an instrument one can do some experiments without the need of a preamplifier (e.g., Experiments 6, 8, and 10). The majority of the less expensive CRO's will have sensitivities of about 10 mV/cm. As long as a good preamplifier is available these less expensive instruments are perfectly adequate, providing other features needed for particular applications are not excluded.

The *frequency response* characteristics of the vertical amplifier are also of great interest. Ideally, for biological applications, the vertical amplifier should be flat from d.c. to several hundred kHz, thus permitting application to practically all recording needs. However, if the instrument is not d.c.-coupled, it can still be used for many experiments. For instance, Experiments 4 through 10 of this manual may all be done with a.c.-coupled amplification, providing good frequency response exists in the lower range. The frequency response characteristics of the horizontal amplifier are of less interest to most biologists, but for some applications requiring input to the horizontal amplifier, obviously the specifications would be of interest; otherwise the horizontal axis is controlled by the sweep circuit.

Just as in the case of the preamplifier, the vertical amplifier of the CRO may also have provision for differential or push-pull input. Whether this is used will depend upon the nature of the input signal: if the signal is fairly large, differential input is less important. Also, the output of a preamplifier may have only single-sided output, in which case single-sided input is used for the CRO. Alternatively, the preamplifier may have differential coupling for both input and output, but the CRO has only single-sided input, consequently, differential input is used for the preamplifier, and single-sided coupling is used for the output of the preamplifier and the input of the CRO. Whenever possible, it is better to use differential coupling throughout.

For certain experiments (e.g., Experiment 10) an *external trigger* input to start the sweep of the CRO is very useful (Fig. 7). The better instruments have external triggering capability, but, again, this is not absolutely vital since most general purpose scopes also have provisions for *external sychronization.* In that way, the input signal may be synchronized with the sweep so that the resulting stationary signal may be studied. This is useful in the study of the EKG (Experiment 6) and the compound action potential from nerve (Experiment 10). Synchronization is unimportant for the study of apparently random spontaneous activity, such as is seen in Experiment 7.

A final consideration for the CRO a as read-out device in electrophysiology is the nature of the phosphor screen. Phosphors are given various ratings or "P" values, which reflect both the color of the phosphorescence and its persistence in time. Long persistence is useful for the visual study of slow phenomena like the EKG. On the other hand, short persistence and a blue color are better for photography of the CRO screen. These are conflicting ideals, and if the CRO is to be of maximum general use, a compromise may be necessary. For example, the P7 phosphor has a two layer screen. Since photography is carried out very often in electrophysiology, specification of phosphor type can be important. An excellent discussion of phosphors can be found in Nastuk (1964), and discussions on the photography of CRO traces can be found in both Nastuk (1964) and Kay (1964). Finally, additional reference material on oscilloscopes is available through commercial manufacturers of these instruments (see Part Four).

After the CRO, perhaps the most important read-out device for biological research is the *graphic recorder.* Such recorders make permanent records on moving paper. The two most common means of writing are with an ink-writing pen or with a pen electrode writing on specifically sensitized paper; movement is achieved through electromagnetic deflection. Since the mechanical components have large inertial masses relative to electrons, the response frequencies are limited. Conventional recorders rarely exceed a few hundred cycles per second. The essentials discussed for the CRO still hold for recorders: this means that the two axes still represent time and voltage. Let us consider some specifications of interest.

The horizontal[1] axis is controlled by a chart drive mechanism. The *chart speed* is variable, and the range of speeds will determine its general usefulness. Most recorders have a wide range of chart speeds, usually given in mm/sec, but, on occasion, unusually high or low speeds are required, and the investigator may consult with the manufacturer for optional chart drives.

As with the preamplifier and CRO, the vertical[1] axis is controlled by an amplifier whose specifications are of interest. Thus, *frequency response,*

---

[1] "Horizontal" and "vertical" are used in a general sense by analogy to the vertically oriented CRT display. Graphic recorders vary so greatly in their construction and modular configurations that the graph may be in any geometrical plane relative to the instrument panel.

sensitivity, and *common mode rejection* values should be known. Further specifications of interest will depend upon the applications.

Although expensive instruments, access to graphic recorders is highly desirable because of their ability to give a permanent read-out record immediately. In contrast, while the photographic records from a CRO camera must be developed after the experiment, the camera will record faithfully the waveform of a fast action potential, while the upper frequency response limitations of the recorder will distort the waveform. Consult Nastuk (1964) and Yanof (1965) for further information on recorders.

Other read-out devices include direct reading meters and digital voltmeters, neither of which is used as commonly as the CRO and recorder for electrophysiological work. Usually such applications are specialized, and beyond the intended scope of this manual. For a good discussion on read-out meters, read Newman (1964). As with all instrumentation, the technology of read-out devices is moving very fast, a factor which, of course, influences recording techniques. For further discussions on read-out devices and recording techniques consult Donaldson (1958), Kay (1964), Nastuk (1964), Newman (1964), and Yanof (1965).

Stimulators. Some of the strategy of neurophysiology research is to stimulate one part of the nervous system and to record its consequence at another site. As the need for precise control of the stimulus grew, electronic stimulators were developed, nowadays being standard items of equipment in electrophysiology laboratories.

Stimulation is carried out either with d.c. or with pulses. When d.c. output is the mode of operation, *voltage* is the only parameter which can be varied; but because d.c. stimulation is used so seldom, it will not be discussed further. The other and more common mode of operation is to stimulate with square wave pulses. Three stimulus parameters are of interest: *voltage*, pulse *duration*, and *frequency* or, more appropriately, *pulse repetition rate*. These parameters describe the nature of the pulse or pulse train, and the specifics of the experiment will determine the exact dial settings (e.g., Experiment 10). However, the instrument must have a capability of varying these three parameters if it is to be of general value for even routine experiments.

To be of maximum value the stimulator must also have provision for *pulse delay*—a variable time interval between a triggering or synchronizing pulse and the stimulus output. The value of such capability lies with the need to displace the signal deflection away from the beginning of the sweep trace, thus facilitating accurate measurement. This is illustrated in Figures 6, 7, and 21.

Since the stimulus signal is large compared with the biological action potential, there is a possibility that a recording electrode will pick up the stimulus, thus introducing a very large input to the amplifier, and so obscuring the biological signal. This difficulty can be corrected by *grounding* the preparation and by using a *stimulus isolation unit* (see Fig. 7). The preparation is grounded between the stimulus leads and the recording electrodes. The stimulus isolation unit is a form of isolation transformer, and its use minimizes interference from a stimulus artifact on the CRO trace. Practically all routine work in electrophysiology which involves electrical stimulation also employs provision for stimulus isolation.

Specific stimulus parameters are outlined in the only experiment in this manual (Experiment 10) involving electrical stimulation. It will be useful for the beginner to feed the output of the stimulator directly into the vertical axis of the CRO and to vary voltage, pulse width, frequency, and delay. Visualization of the stimulus output before doing experiments will help in the understanding of how the stimulator works. Additional information on electronic stimulators and stimulation may be found in Donaldson (1958), Kay (1964), Nastuk (1964), Yanof (1965), and Florey (1966).

Transducers. Very broadly defined, a transducer is a device by which energy is transmitted, usually in a different form, from one system to another. In the science of instrumentation transducers are classified in various ways. The classification scheme favored by the author is that of Lion (1959) which can be listed as follows:

Input Transducers, in which a non-electrical magnitude is converted to an electrical output,

Modifiers, in which one electrical magnitude is converted to another modified electrical magnitude, and

Output Transducers, in which an electrical magnitude as input is converted into a non-electrical output.

In electrophysiological instrumentation the term transducer has come to mean input transducer as defined above. Examples of input transducers would include strain gauges, photoelectric cells, piezoelectric pickup devices, etc. Modifiers and output transducers also have applications to biomedical research, although they are not always thought of as being transducers. For example, an amplifier is a modifier and a loudspeaker is an output transducer. Transducer also has another meaning in electrophysiology, especially in describing sensory systems. For example, when light energy impinges upon an eye, biological transducer action is responsible for conversion of the light quanta into electrical signals, while another type of transducer converts mechanical energy to electrical energy. This is discussed in Chapter 6.

While this manual does not include experiments involving the use of input transducers, the use of modifiers and output transducers, and the concept of biological transducers, are a necessary part of the experiments. For that reason, and in the interest of completeness, this section on transducers has been included in the manual. For experiments more complex than those described here, the investigator may require the use of input transducers. Lion (1959) is an excellent source of material on input transducers; Goldstein (1964) describes simple input transducers suitable for teaching; the general applications of transducers to biomedical research are discussed by Brown and Saucer (1958), Donaldson (1958), Nastuk (1964), Yanof (1965), and Rushmer (1966).

Chapter Three

Instrumentation Systems

The various components are put together into what may be called an instrumentation system. At this point it might be wise to divert ourselves briefly in order to consider some basic premises. The primary consideration when building up an instrumentation system is the type and scope of experiments envisaged; never should it be attempted to fit in some experiments with apparatus which happens to be available, or which might have been acquired in some haphazard fashion. As is discussed further in Appendix 2, the biological experiment is of paramount importance, either as a teaching exercise or a research activity. At any rate, we will proceed on the premise that having decided what experiment is to be done, the instrumentation system is assembled accordingly. The conception becomes complete; only the physical assembly remains.

Let us start by saying we wish to stimulate and record from isolated nerves as described in Experiment 10. To study the compound action potential we will need a stimulator, amplifier, and CRO as minimum equipment. In addition, we will need a nerve chamber, electrodes, hook up wire, etc. In the paragraphs that follow, three instrumentation approaches to studying the same experiment will be described. This will emphasize that the solution to instrumentation needs can be found in many ways. Up to now, the mention of any instrument manufacturer has been avoided. Now it will be necessary to mention specific instruments.

Method I. The basic instrument requirements for the first method are supplied by the Heath Model EVW-3 IMPScope. The IMPScope is three instruments in one; it consists of stimulator, amplifier, and CRO. It is, in fact, an instrumentation system rather than a single instrument. It is the least expensive and simplest to use of the commercially available possibilities for the performance of electrophysiological experiments. A diagram of the experimental

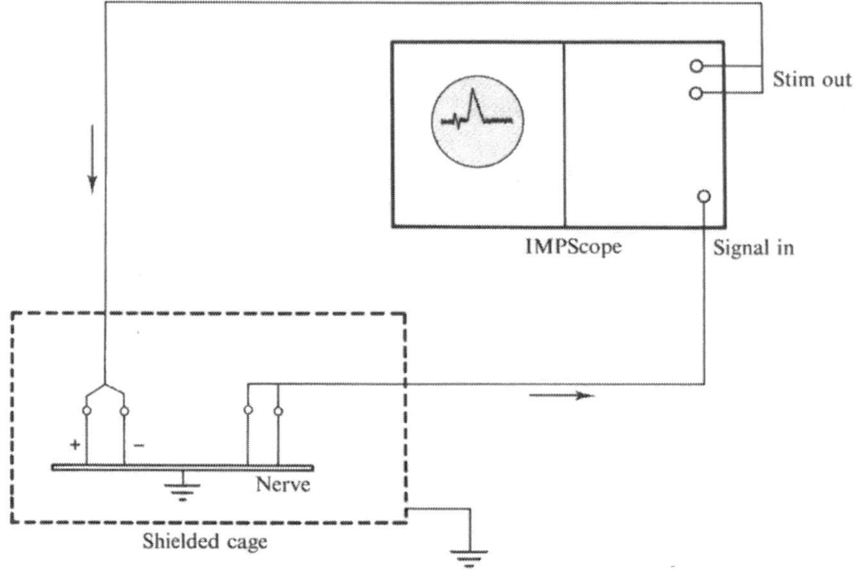

FIG. 5. Diagram of the experimental arrangement to study compound action potentials in nerve using the Heath Model EVW-3 IMPScope (Heath Company, Benton Harbor, Michigan 49022).

setup using the IMPScope is shown in Fig. 5. The nerve preparation is placed inside a *shielded cage*, which consists of copper or galvanized wire screening covering a simple wooden box frame on all surfaces except one side through which access is available to the inside. All surfaces covered with screening must be in good electrical contact. Convenient, but not critical, dimensions for the cage are $24'' \times 24'' \times 36''$. With good push-pull amplification and shielded input cable, a cage may not be necessary. Nevertheless it is an excellent precaution against interference, and the cost of construction is minimal compared with the electronic apparatus.

Note the minimal need for wiring connections with this method. Two stimulating leads go to one end of the nerve, while two recording leads are connected at the other end. Note the position of the cathode. *Ground* connections are made to the nerve, cage, and instrument. These ground connections are general requirements for all instrumentation systems. The physical details of how grounding is achieved depend upon the connecting cables, availability of ground terminals on the chassis, and whether the power cord has provision for safety grounding in a 3-prong plug. If an instrument is grounded through a power cord, other units may be connected to the instrument for grounding, but if no such ground is available, the best way to establish one is to make a good electrical connection with a cold water pipe. Ground connections should be made at one terminal, preferably on the chassis of the CRO if it is grounded through the power cord, or directly to the ground point if not so grounded. Multiple grounds from any component may result in *ground loops* and an increase in interference. Improper grounding can be a major problem for beginners in electrophysiology. If problems are encountered, persistence and experimentation with instrument positions and connections is necessary; at any rate, the experimenter should not be discouraged easily. Additional hints on connections, shielding, and proper grounding can be found in Whitfield (1953), Suckling (1961), Nastuk (1964), and Yanof (1965).

Method II. Another instrumentation system for recording nerve action potentials is shown in Fig. 6. The Grass Model SD5 Stimulator is supplied with a built-in stimulus isolation unit. The stimulus output goes to the nerve while a synchronizing signal goes to the CRO. The nerve action potential is fed by differential input coupling to a Grass Model P9B Preamplifier. Note that since the EICO Model 427 CRO has no push-pull input to the vertical amplifier, the one-sided output from the P9B is used. Power is supplied to the preamplifier through a Grass Model RPS109 Power Supply. The amplifier being inside the

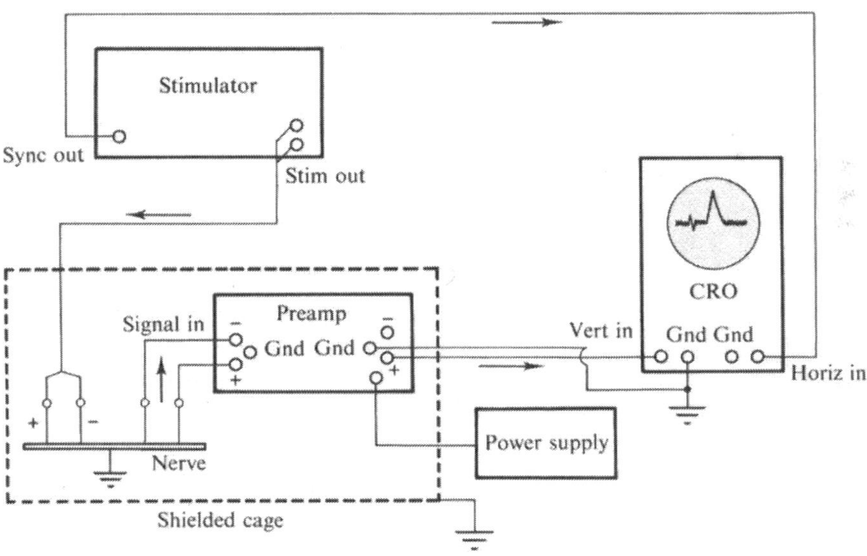

FIG. 6. Block diagram of alternative apparatus to study the compound action potential. Stimulator model SD5, Preamplifier Model P9B, and Power Supply Model RPS109 are from Grass Instrument Company (Quincy, Mass. 02170). CRO is an EICO Model 427 (EICO Electronic Instrument Company, Inc., Flushing, New York 11354).

shielded cage, noise interference is reduced. The stimulator, and the preamplifier and its power supply, are grounded through their power cords. The EICO oscilloscope must be connected to a ground, as must the nerve and the cage. The synchronizing signal from the stimulator is fed into the horizontal input terminal of the CRO. If the CRO had provision for external triggering (see Method III), this synchronizing signal would feed into the external trigger input terminal.

Method III. The third method of studying the action potentials in nerve is exactly the same in principle as the previous methods. The instrumentation system is illustrated in Fig. 7. The stimulus is supplied by units of the Tektronix Type 160 Series connected together as shown. A Bioelectric ISA Isolator is used as a stimulus isolation unit. The nerve action potential is fed by differential input into a Tektronix Type 1A7A High Gain Amplifier plug-in unit which in turn is plugged into a type 547 CRO. This instrument will accept many different

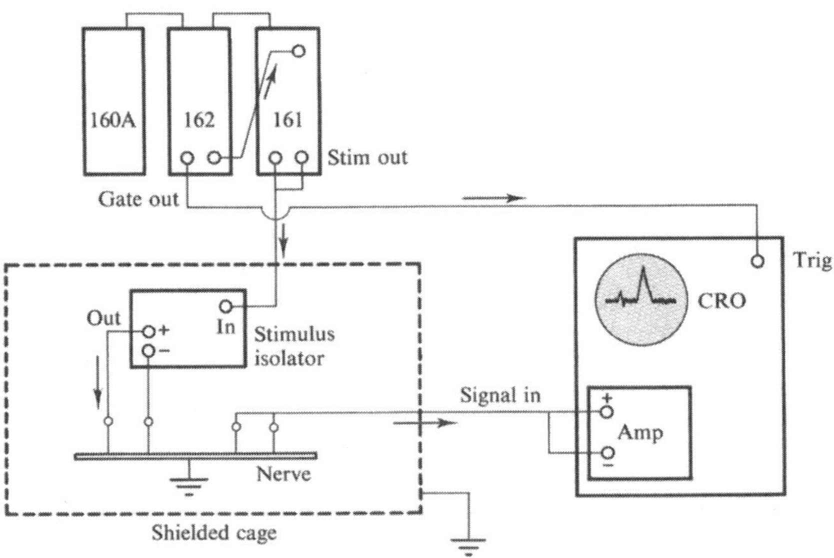

FIG. 7. Block diagram of Method III for studying action potentials in peripheral nerve. The following Tektronix (Tektronix, Inc., Beaverton, Oregon 97005) units are used:

Type 160A   Power Supply   Type 161   Pulse Generator
Type 162   Waveform Generator   Type 1A7A   High Gain Amplifier
Type 547   CRO

The stimulus isolation unit is a Bioelectric ISA Isolator (Bioelectric Instruments, Inc., Hasting-on-Hudson, New York, 10706).

types of plug-in units, allowing for many applications. An output gating signal from the Type 162 Waveform Generator goes to the trigger input terminal of the CRO. The delay between the gating signal and the stimulus pulse is adjusted by an appropriate setting on the output pulse delay dial of the Type 161 Pulse Generator. Frequency is adjusted on the waveform duration dial of the Type 162. Stimulus voltage may be controlled both at the Type 161 and at the stimulus isolation unit. Grounding of the Type 160A and Type 547 takes place through the 3-conductor power cords. As with the previous methods, proper grounding of the cage and nerve must be carried out, however in this specific case the nerve

TABLE 2. COMPARISON OF INSTRUMENTATION COSTS FOR THREE ALTERNATIVE METHODS OF STUDYING THE COMPOUND ACTION POTENTIAL FROM PERIPHERAL NERVE. PRICES ARE SUBJECT TO CHANGE AND MAY NOT NECESSARILY BE CURRENT.

Method	Components	Approximate component cost in dollars	Total system cost in dollars
I	*Stimulation*		
	Stimulator section of Heath EVW-3	—	
	Amplification		
	Amplifier section of EVW-3	—	
	Display		
	CRO section of Heath EVW-3	—	280 (Assembled)
II	*Stimulation*		
	Grass SD5	250	
	Amplification		
	Grass P9B	140	
	Grass RPS109	175	
	Display		
	EICO 427 CRO	140 (Assembled)	705
III	*Stimulation*		
	Tektronix Type 160A	215	
	Tektronix Type 161	145	
	Tektronix Type 162	145	
	Bioelectric ISA Isolator	200	
	Amplification		
	Tektronix Type 1A7A	440	
	Display		
	Tektronix Type 547 CRO	1875	3020

could be connected to the cage and the cage connected to a ground terminal on the CRO.

The three methods discussed above are only examples selected from many possible instrument combinations. They do not necessarily represent the best selection nor by their omissions do they suggest that other manufacturers produce less suitable instruments. The specifics of how the actual experiments are done can be found in Experiment 10. The instrumentation units and their costs for the three methods are compared in Table 2. It should be clear that cost figures must enter into selection. More information on the selection of specific apparatus may be found in Appendix 2.

The emphasis of this section has been on how various components can be assembled into instrumentation systems and how they relate to the performance of specific experiments. Additional diagrams and applications of instrumentation systems to electrophysiological experiments may be found in Welsh and Smith (1960), Suckling (1961), Goldstein (1964), Whitfield (1964), Florey (1966), Bures *et al.* (1967), and Hoff and Geddes (1967). In addition, a manual published by the American Physiological Society (1967) has a good section on electrical phenomena in excitable tissues.

Before ending this section it should be emphasized that much more elaborate instrumentation systems are used for serious research in neurophysiology. While such systems are beyond the scope of this simple manual, it might be appropriate to mention some of the instrumentation. Figure 8 is an attempt to summarize how additional units might fit into the basic scheme already described. Note how an audio system could be used, and how a camera and graphic recorder could be used to make permanent records. Such units are employed in routine fashion, and specific references to their possible use are mentioned in some of the experiments in this manual.

Let us consider now that part of the figure shown as dotted lines, which represent operations and signal flows that are usually not routine, at least not for teaching purposes, and so unlikely to be encountered by the beginner. Nevertheless they are added in this manual so that the reader might gain slightly greater perspective on various applications. It should be remembered that, in addition to *detection* of biological signals, *analysis* is an important function. The various instruments may also be called *signal conditioners*. By using analog to digital (A/D) converters as interface equipment, it is even possible to feed the various signals into a high speed digital computer programmed for specific analyses. The advantage, of course, would come from the enormous saving in time. However, computers and even other special applications devices are usually expensive and, consequently, are not used extensively. Consult Tolles (1964b) for various computer applications in biology and medicine.

Note also in the figure one final signal flow line, the feedback signal. An appropriate output from a computer or other instrument could be made to feed back to the stimulator. This feedback signal could then control the stimulator

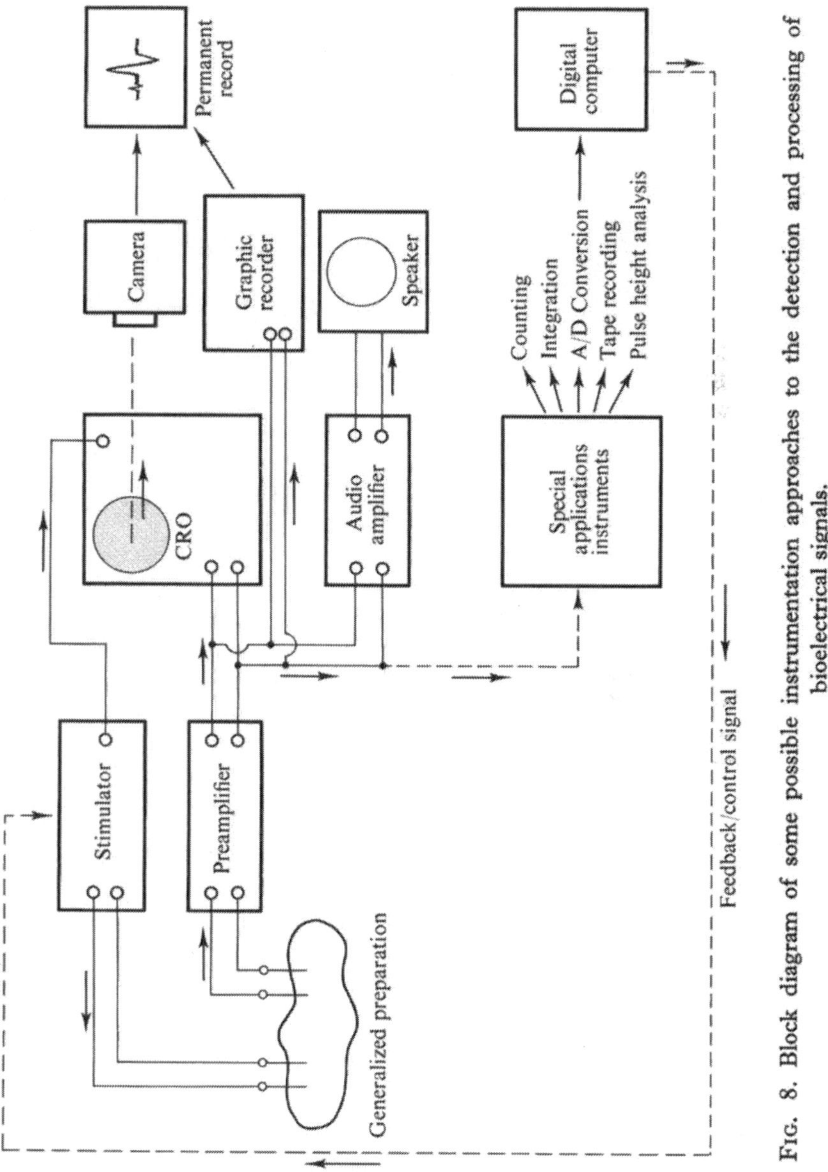

FIG. 8. Block diagram of some possible instrumentation approaches to the detection and processing of bioelectrical signals.

to perform a specific stimulus. The bioelectrical output then goes through the system again and, in turn, some function of it returns to the stimulator, and so on. This now is a *closed loop* system from biological preparation to instrumentation back to biological system. Complex biological phenomena are now being studied in this manner. Such studies will provide one of the more exciting chapters in electrophysiological research in the future.

Part Three

LABORATORY EXPERIMENTS

Chapter Four

Electricity: Basic Phenomena

It is both interesting and significant in the history of science that the fields of electrochemistry and electrophysiology began about the same time. The great controversy of Volta and Galvani was the basis of experimentation that eventually established both of these sciences on a firm footing. The reader is urged to study the work of these two great investigators.

In some of his early experiments Galvani hung frogs' legs by bronze hooks that were, in turn, suspended from an iron railing. Whenever the frog tissues completed the circuit—frog to iron to bronze to frog—the legs twitched. We now know that the basis of this was the irritable tissues responding to the potential difference generated by the dissimilar metals in contact with the moist tissues. In later years the frog leg muscle became a sensitive instrument to detect current. Also in later years, many layers of dissimilar metals were arranged consecutively along a forceps-like arrangement for use as a stimulator in electrophysiological studies. When the investigator was ready to stimulate, he soaked the stimulator in weak acid solution; the d.c. that was generated could stimulate the biological preparation. In the following experiments, both the generation of current by dissimilar metals, and the use of a frog as a current detector, will be demonstrated.

Electromotive Force:
Dissimilar Metals

For the first experiment, a voltmeter or a CRO will satisfy our instrumentation needs. The sensitivity need not be high; 1–10 volts full scale on the voltmeter, and about 0.1 to 1.0 V/cm on the CRO are sufficient. Next, obtain a penny (Cu), a nickel (Ni), a dime (Ag), and a piece of zinc (Zn) plate. Place a piece of filter paper or ordinary paper towel moistened with a weak saline solution between any two of the metals. Connect one metal to one terminal of the recording instrument, and the other metal to the second terminal. If a CRO is used, it must be d.c.-coupled. A switch may be placed in series in the circuit (Fig. 9). Determine

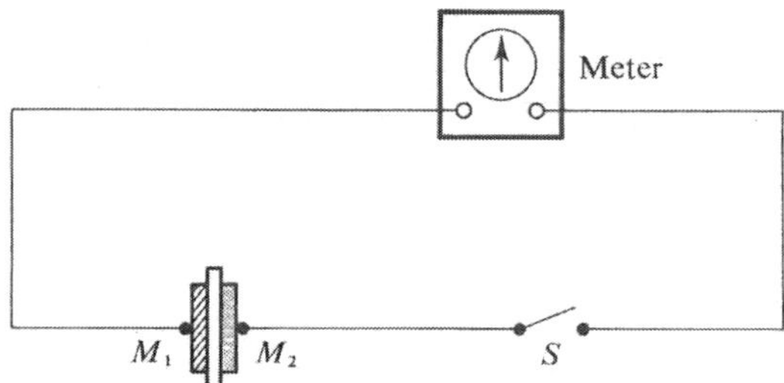

FIG. 9. Diagram of the circuit arrangement to study the electromotive force generated by dissimilar metals (M_1 and M_2) separated by a piece of paper saturated with saline.

the potential difference. Which metal is positive? In a similar manner, measure the potential differences and polarities of all the other metal combinations. Enter the data in an appropriate table (e.g., Table 3). Discuss your findings in the light of electrochemical data from a physical chemistry book or other reference.

TABLE 3. TABLE FOR ENTRY OF DATA FROM THE
EXPERIMENTS WITH DISSIMILAR METALS

Metal pair	Voltage in mV	Metal forming positive pole
Cu-Ni		
Cu-Ag		
Cu-Zn		
Ag-Ni		
Ag-Zn		
Ni-Zn		

Galvani's Experiment

Now we can turn our attention to a living organism, the frog. Decapitate a frog and remove the forelimbs, viscera, and skin from the remaining backbone and hindlegs. Suspend the carcass of the frog with a hook made out of stout copper wire; the hook should go through the backbone and the spinal canal (Fig. 10). Next, bring a zinc plate or strip in contact with one foot. The zinc, in turn, should be attached with a metallic clamp to a stand from which the copper hook is

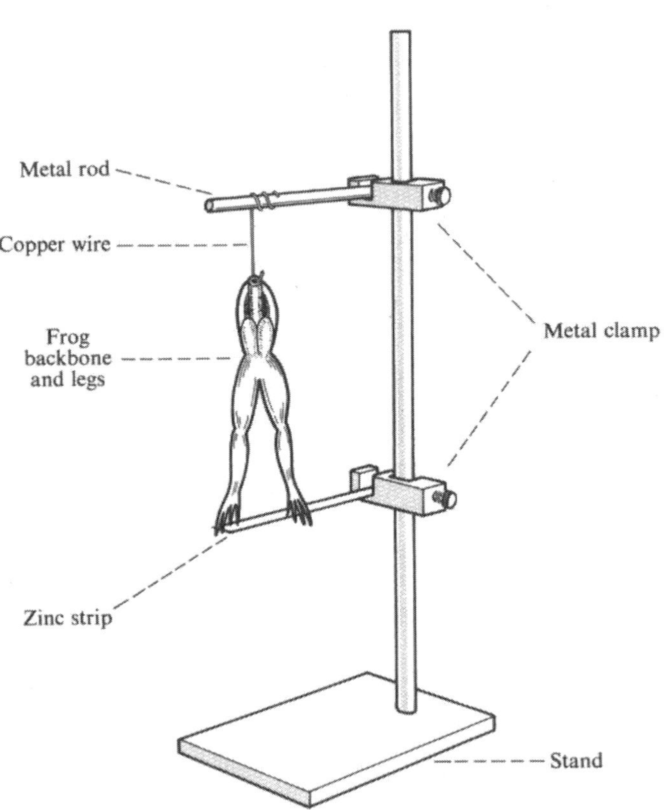

FIG. 10. Diagram of the experimental arrangement for the duplication of Galvani's Experiment.

hanging. It is important that all contacts are clean so that a complete circuit exists from stand to copper wire to frog to zinc to stand. Describe the results.

A simpler version of this experiment can be carried out with a manual bimetallic nerve stimulator, made from a piece of copper wire (about 20 gauge) and a piece of zinc wire of about the same diameter (if zinc wire is unavailable, a practical alternative would be thin sheet zinc cut into strips 1–2 mm wide). Twist together the Cu and Zn wires or strips about 6 to 8 cm long so that they make good electrical contact. Expose the spinal nerves (numbers 7, 8, or 9) going to the hindlimbs of a frog. At about 1 cm apart, apply the free ends of the stimulator to one of the exposed nerves. What happens when you touch the two wires to the nerve? Make a stimulator out of two copper wires and test it in the same way. What happens?

In these experiments we generate electrical potential by two dissimilar metals. The frog serves two purposes. First it supplies the moist tissues which together with the metals generate the electrical potential, and second the frog nerve and muscle tissues respond to these electrical currents so that, in effect, they serve as a recording instrument. As a matter of fact, the frog nerve-muscle preparation known as a *rheoscopic frog* served as an electrical measuring instrument during the early part of the nineteenth century, before sensitive galvanometers were available. The reader is directed to Green (1953), Brazier (1959), and Galambos (1962) for additional information of Galvani and the early beginnings of electrophysiology.

Demarcation Potentials in Muscle

Pith a frog and carefully remove the two gastrocnemius muscles. Place the muscles in frog saline (Ringer's Solution; see Appendix 1) and leave them there until you are ready to record the potentials. Prepare the apparatus so that the muscle, a sensitive galvanometer, and a switch are all connected in series (Fig. 11). Cut across the belly of one muscle somewhat near the femur end and

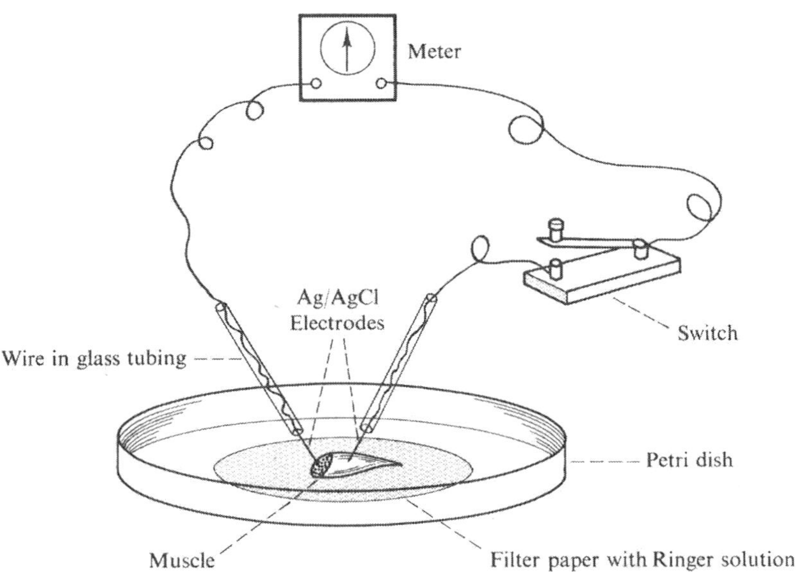

Fig. 11. Diagram of the apparatus and arrangement used to demonstrate a demarcation potential from an isolated frog muscle.

connect it to non-polarizable electrodes (Ag/AgCl or wick electrodes). One electrode is connected to the intact muscle surface and the other on the cut surface. Keep the muscle moistened on a piece of filter paper. However, be careful that too much saline doesn't short out the current. Close the switch and note the direction and magnitude of deflection of the galvanometer. Change the distances of the electrodes relative to each other and note the magnitudes of the resulting deflections. Reverse the positions of the electrodes and note the direction of deflection. It is possible to arrange two cut muscles in series and thus increase the magnitude of the recorded potential. Again, care should be taken to avoid shorting out.

Various chemical substances may affect the demarcation potential. Try immersing a muscle in isotonic KCl solution. Test any other substances the laboratory instructor may suggest. For additional information on demarcation (or injury) potentials in muscle, consult Suckling (1961), Galambos (1962), Florey (1966), and Tasaki (1968).

Chapter Five

Bioelectrical Action Potentials

In this section we will examine bioelectrical potentials generated by various tissues from various animals. Moreover, all will be examples of phasic potentials, representing excitable tissues in a state of activity. The experiments have been selected to remind the reader that the occurrence of bioelectrical potentials is widespread, both in terms of animal species and in types of tissues and organs. The reader is encouraged to extend this list of experiments by experimenting with heart, muscle, and nerve tissues from species not included in this section.

Electric Organ Discharges

The powerful electric discharges of the electric eels and electric torpedoes have been known since antiquity. In fact, the ancient Greeks called the torpedo *Narce*, meaning "that which numbs." This word is the base of our English word "narcotic."

Even though the powerful shocking effects were known early, they were not identified as coming from electricity until the eighteenth century, when it was finally recognized that animal electricity and physical (or atmospheric) electricity have a common basis. The anatomists of the nineteenth century noted certain organs in some fishes that resembled the electric organs of the well-known electric eels of South America. Since these fishes did not give powerful discharges, the organs were called pseudoelectric organs. Then it was discovered that these fishes do, indeed, generate weak electrical discharges. There are at least nine families of fishes that can be classed conveniently as weakly electric fishes. These discharges are not powerful enough to shock a person, but sufficiently powerful (a few volts) to be detected with simple instrumentation.

For purposes of simplicity we can classify the weakly electric fishes into two groups: (1) the pulsating fishes (e.g., *Gymnotus, Hypopomus*) and (2) the a.c. fish (e.g., *Sternarchus, Sternopygus, Eigenmannia*). These fishes can be obtained frequently from commercial tropical fish dealers, who know them as "knife-fishes." The fish may be kept in clean tanks, preferably not in too cold a room. Details on feeding, water treatment, etc., can be supplied by the dealer.

Place one of these fishes in a small glass or plastic container filled with distilled water. The fish will suffer no harmful effects from the distilled water, and its electrical signals will be considerably less attenuated. Then place two metallic or carbon rod electrodes into the container, generally at opposite ends (Fig. 12). The electrodes are then coupled by alligator clips and wire leads directly into a CRO. The sensitivity may be set at about 0.5 V/cm. An audio unit will greatly aid in the study of the electrical pulsing pattern. Usually an audio amplifier will be required to provide sufficient power for a suitable loudspeaker. Any simple audio amplifier and associated speaker will serve with this preparation.

Determine whether the fish you have is a pulsating or a.c. type. Determine also its frequency of firing and the maximum amplitude of the electrical signals. Note, expecially, the marked changes in amplitude as the fish moves, thus changing the geometry between fish and recording electrodes.

Study the effects of temperature change, to which a.c. fishes are especially sensitive, on the frequency of discharge.

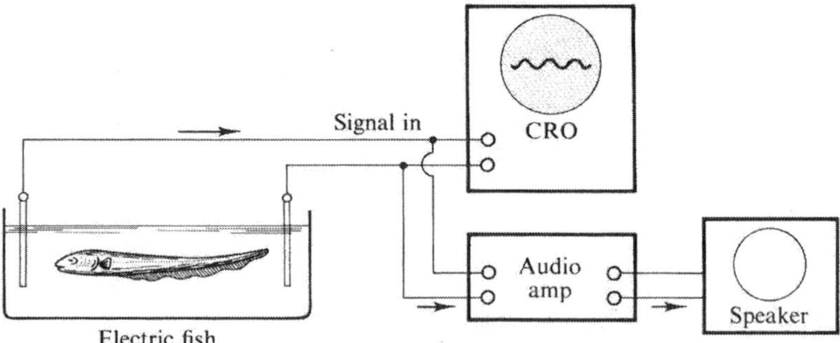

FIG. 12. Block diagram of the apparatus used in studying electrical discharges from weakly electric fish.

Using a small battery and another set of electrodes, try introducing weak electric pulses into the water. The pulsating fishes often change their discharge rate when they sense "foreign" electrical pulses.

The electric organ is similar to a dipole in the body of the fish. Try to restrain the fish by allowing it to swim in a tube made of plastic screening, then move the electrodes systematically relative to the body axis of the fish, and determine the electric field lines. Isopotential lines can be determined readily.

Besides having electric organs these fishes have sensitive electroreceptors. It is thought that they can navigate underwater by detecting small changes in the electric fields which they generate. It is possible to study the sensitivity of the electroreceptors by making use of a conditioned response. After the fish is conditioned to associate food with an a.c. signal, its sensitivity may be determined by reducing the signal systematically at subsequent trials over a period of time.

Lissmann (1958) and Grundfest (1960) provide general information on electric organs. Lissmann (1963) explains how electric organs of weakly electric fishes are thought to aid in underwater navigation. An advanced monograph on electric organs has been edited by Chagas and Paes de Carvalho (1961). Additional information on electroreception thresholds and electrical fields in weakly electric fishes may be found in Hainsworth *et al.* (1963), and Granath *et al.* (1967, 1968).

Muscle Action Potentials

Muscles from a variety of animals generate action potentials as they contract. These action potentials are electrical signals which travel along the muscle fibers, weaker signals, however, than the electric organ discharges, consequently, needing amplification. In this experiment, therefore, we must make use of additional amplification.

We can use very similar techniques in this experiment. A large finger bowl, crystallizing dish, or porcelain dish about 15–20 cm in diameter, and deep enough to allow a water depth of several centimeters, will serve as a suitable container for distilled water. The remote electrodes (heavy copper wire or graphite rodes) are lowered perpendicularly into the water at a distance of about 12–15 cm apart. The upper ends of the electrodes are clamped by alligator clips which are connected to electrical leads which are then fed into the input of a preamplifier. Precaution against pickup of electrical artifacts is necessary when sensitive amplifiers are used, therefore, the container and preamplifier are placed inside a shielded cage. The output of the preamplifier is fed in parallel into a CRO and the audio system. As with the previous experiment, an audio amplifier and loudspeaker will help in studying the action potentials. The instrumentation system is shown in Fig. 13.

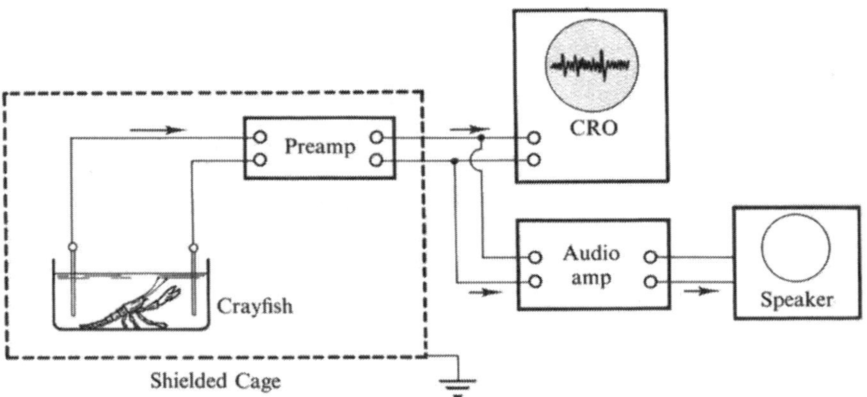

FIG. 13. Block diagram of the method and equipment used to detect and study muscle action potentials from animals in an aquatic medium.

A good-sized crayfish (3–4 inches long) will make an excellent experimental animal, which may now be placed within the container, and the instrumentation system will record the bioelectrical potentials generated by the body of the animal as it moves about. The relatively slow walking movements can be correlated easily with the electrical signals as monitored by the CRO and audio system. As the animal remains relatively still, the baseline of the CRO is steady; as it moves, muscle action potentials with amplitudes in the order of 50 to several hundred microvolts are picked up. The sensitivity of the recording system should be set to give about 50 to 100 μV/cm deflection. The bioelectrical potentials are generated by the muscles and are conducted throughout the volume of the water. The amplitudes are greatest when the animal is closest to the recording electrodes. If the surface of the water is disturbed, the resulting waves will produce corresponding, slow oscillations of the baseline of the CRO. However, these disturbances should be minimal and should not present a problem.

Bioelectrical potentials accompanying the movements of a variety of animals may be recorded by this method. These include frogs, fish, and aquatic insects. Even the EKG from a frog may be picked up, provided that the more pronounced muscle action potentials do not mask it. However, the EKG is better recorded with contact electrodes, as will be described in the next experiment. The method of remote electrodes may also pick up human muscle action potentials: the hand is placed in the water between the electrodes: placing the other hand on the shielded cage will help ground out possible 60 Hz artifacts, then, on alternately clenching and relaxing the hand as a fist, the potentials can be detected dramatically.

While the method of remote electrodes is simple, it can pick up gross muscle action potentials only, and cannot localize the active tissues. For those wishing to record directly from muscle, the following experiment may be tried. Restrain a crayfish with rubber bands on a small piece of wood, dorsal side up. Grasp one walking leg and with a needle puncture a small hole in the antero-dorsal surface of the membrane between the carpus and propus. Insert a fine, insulated copper wire (e.g., 34 gauge) into this hole until about 1 cm of its length lies inside and parallel to the long axis of the propus. Using an alligator clip, clamp the other end of this flexible wire and feed into the preamplifier. Place the crayfish in the container of water. Another wire placed anywhere in the water serves as an indifferent electrode. The crayfish is free to move because of the flexible wire, and the muscle action potentials are recorded by the very tip of the wire where the insulation is cut. Depending upon the precise location of the wire, it is possible to pick up both nerve and muscle action potentials; however, the muscle action potentials should predominate.

With either method of recording, study the pattern of action potentials from the animals, both at rest and while walking. With the remote electrode method, place the animal in a weak saline solution instead of distilled water. Explain the results. Repeat this experiment while employing the flexible wire electrode in the leg. Again, explain the results. Crayfish are sensitive to sudden light changes, and experiments with sudden changes in ambient light can be carried out to good advantage. In a similar manner, determine what other physical, environmental changes might influence the behavior of the crayfish.

Additional information on the remote electrodes method and its application to recording from aquatic animals can be found in Camougis (1960a, 1960b). Consult appropriate sections in Suckling (1961), Florey (1966), and Katz (1966) for discussions on bioelectrical potentials associated with muscle tissues.

Experiment 6

Electrocardiograms

The electrocardiograms (EKG's) of various animals make excellent subjects for experimental study. The EKG potentials are relatively large and periodic in nature, thus permitting easy detection; the periodicity of the heart beat lends itself to experimentation with factors influencing heart rate; and, finally, the sensitivity of the heart to drugs makes it an excellent preparation to study the actions of drugs on excitable systems.

The essentials of recording an EKG are to attach electrodes to the body of any animal, thus picking up the relatively large bioelectrical fluctuations which accompany the heartbeat—these fluctuations in potential are conducted throughout the volume of the body so that electrodes need not be on the heart itself. It is this principle that permits clinical electrocardiography in human medicine. However, if the electrodes are on the heart itself, the detection of the EKG potentials is, of course, much easier. In this section we will consider experiments with a mammal (rat), a cold-blooded vertebrate (frog), and an invertebrate (crayfish).

Anesthetize a medium sized rat (150 to 250 grams) with sodium pentobarbital (40 mg/kg) or α-chloralose (55 mg/kg) injected intraperitoneally. After the animal is anesthetized, clip the four limbs to an appropriate animal board. Syringe needles are used as electrodes (the size is not critical; 1 inch, 26 gauge needles are quite satisfactory). Insert the needles subcutaneously on either side of the rib cage, ensuring good electrical contact with the tissues. Banana plugs or alligator clips will serve as connectors to the leads going to the preamplifier.

Pith or anesthetize a frog and pin it, ventral side up, on a frog board or dissecting tray. Cut open the abdominal and thoracic cavities so as to expose the heart. The heartbeat will continue for hours provided trauma and bleeding are minimized and the heart is kept moistened with frog Ringer solution. Place a metal electrode in contact with the ventricle, but positioned so that no mechanical injury occurs to the beating heart. (Some investigators prefer a cotton wick electrode on the heart to improve contact and prevent injury.) Another electrode placed on exposed body wall tissue will serve as a suitable indifferent electrode. The two electrodes are then coupled to leads as an input to the recording system.

The principles of recording the EKG from the crayfish are very similar. The heart lies just under the carapace in the middle of the posterior region of the cephalothorax. Restrain a medium-to-large sized crayfish by clamping it around the cephalothorax or by holding it on a small board with rubber bands. The dorsal posterior part of the carapace is then carefully removed, thus exposing the heart. Place a metallic or cotton wick electrode directly on the dorsal surface of the heart. Sometimes a small ball of solder or silver at the end of the electrode will prevent mechanical injury to the heart and improve electrical contact. Again, the indifferent electrode may be placed on any convenient exposed tissue. The preparation is coupled to the instrumentation system in the conventional manner.

Let us now turn to the instrumentation system. The same recording system may be used for all preparations (Fig. 14). The signals are fed into a preamplifier. The output is fed into a CRO. The sensitivity of the system should be set at about 0.5 to 1.0 mV/cm. The sweep should be slow. If the CRO is equipped with an external trigger, the signal may also be used to trigger the sweep. Otherwise the sweep could be synchronized with the heart rate. This will permit the viewing of one to several EKG signals on the screen as stationary waves, thus permitting better observation of the waveform. If available, a graphic recorder should be connected in parallel with the CRO: it will be found very useful for studying the EKG, since changes in waveform, amplitude, and frequency are permanently recorded and easily studied.

After the animal is set up for recording, note the waveform, amplitude, and frequency of the EKG. Make visual correlations between the EKG and mechanical contractions of the heart in the case of the frog and crayfish. With the rat, the heartbeat can be detected by palpation with the fingers. Several experimental procedures may be pursued. For example, study the action of different temperatures on the frog heart. Temperature may be altered by slowly dripping Ringer solutions of low (15°C) or high (35°C) temperature on the heart. Study

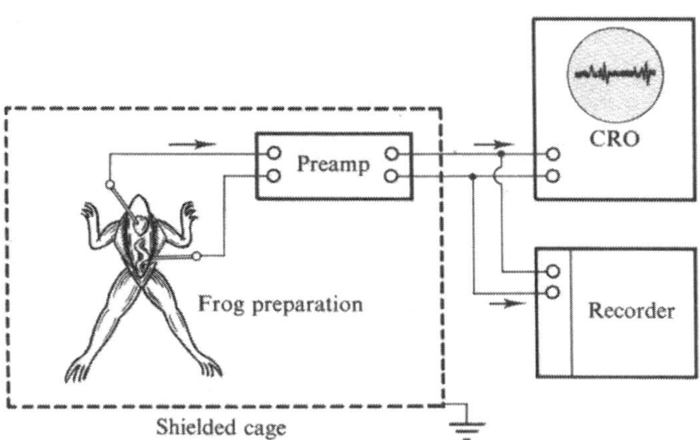

FIG. 14. Diagram of a technique to record the EKG from a frog. The active electrode is on the exposed heart while the indifferent electrode is on abdominal body wall tissue.

the actions of tincture of digitalis, adrenaline, and acetylcholine on the heart. Determine which concentrations, routes of administration, and on which species these drugs are active. (Adrenaline and acetylcholine being unstable in solution, fresh solutions must be used for each experimental day.) By placing the needle electrodes on different limb combinations in the rat, convince yourself that the electrical currents associated with the EKG flow throughout the volume conductor of the three-dimensional animal body. Note, also, that the waveform changes with electrode positions. For this reason, the lead positions are standardized for cardiovascular research with animals and for clinical electrocardiography.

Biophysical principles underlying electrocardiography are discussed by Bayley (1958). Additional references include a conference monograph (Hecht, 1957), and two textbooks (Hoffman and Cranefield, 1960; Suckling, 1961).

Experiment 7

Nerve Action Potentials

In the previous experiments attention was centered on the bioelectrical potentials from electric organs, muscle, and cardiac tissues while in a state of activity. Nerve tissue in a state of activity will also produce action potentials. The classical way to study nerve action potentials has been to stimulate and record from a section of isolated peripheral nerve, usually from the frog (e.g., Experiment 10). However, it is also possible to study action potentials from an isolated, simple, central nervous system preparation. Arthropods offer excellent possibilities to study such nerve action potentials.

In this experiment the preparation will consist of the abdominal portion of the ventral ganglia of the crayfish. Cut off the abdomen from a medium-to-large crayfish as close to the cephalothorax as possible. Pin out the abdomen flat, ventral side up. Remove a strip of integument about 5 mm wide from the median ventral surface, thus exposing the abdominal portion of the ventral nerve cord of ganglia. Grasp the anterior end of the ganglia with fine forceps and gently pull the ganglia away from the surrounding tissue. Use fine scissors to cut connecting fine nerves if necessary. Usually it will be necessary to sever the many nerves radiating from the last ganglion before the chain is completely isolated. Place the isolated ganglia immediately in crayfish saline (van Harreveld solution; see Appendix 1). As with all isolated tissues, precautions must be taken to prevent drying out during isolation and experimentation procedures. A common mistake with beginners is to underestimate how quickly a piece of small tissue will dry out when exposed to room conditions. It is the author's experience that this point must be stressed repeatedly.

After isolation, the ganglia are ready for mounting on recording electrodes. Place the ganglia gently across fine silver wire electrodes in an appropriate nerve chamber. Since the ganglia are spontaneously active, it is not necessary to employ electrical stimulation. Connect the electrodes to the amplifying and recording system (Fig. 15). Note the constant discharge of nerve action potentials of various amplitudes and waveforms. An audio system is particularly helpful in the study of such discharges, as the ear will often detect changes in activity which the eye cannot detect on the CRO screen. Set the amplification so that the sensitivity of the system is about 50 μV/cm on the CRO. The sweep speed should be relatively slow; 10 msec/cm or a sweep frequency of about 10/sec is satisfactory. It may be necessary to increase sweep speed if variations in waveform are to be studied in more detail.

With the help of the calibration reference of the recording system, determine the amplitudes of the various spikes. Also, by means of the time base,

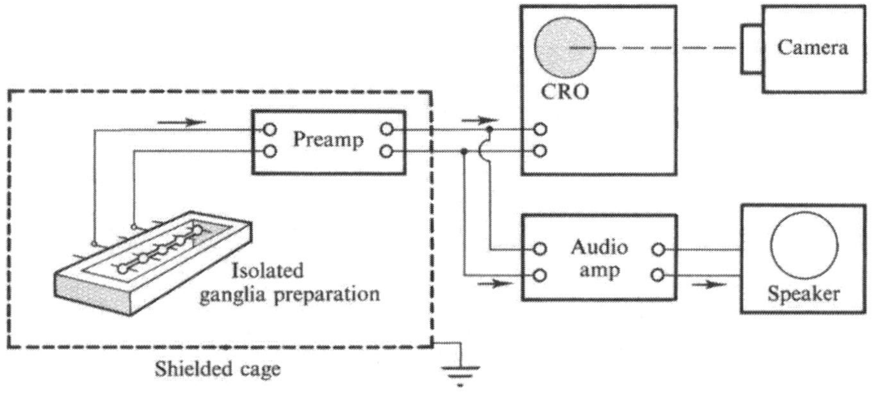

FIG. 15. Block diagram of the electrophysiological equipment for studying spontaneous electrical activity from the isolated, abdominal ganglia of the crayfish.

estimate the approximate frequency of discharge of the spontaneous activity. The central synapses in the ganglia are very sensitive to a variety of drugs. Crayfish ganglia are especially sensitive to nicotine. Gently blow one or two puffs of tobacco smoke on the ganglia and note any changes in the frequency of firing. Some filter brands may produce no results; while on the other hand, too much nicotine may cause irreversible block of electrical activity.

One of the remarkable properties of the terminal ganglion of the crayfish nerve cord is its photosensitivity. An increase of light will excite the ganglia to fire at higher frequencies: this response lends itself nicely to study. Reduce the light of the room and allow dark adaptation for a few minutes. Then shine a flashlight directly on the ganglia and note the response. If appropriate filters are available, the spectral sensitivity may be studied. If a kymograph camera is available, the permanent photographic records of spike discharges will greatly facilitate the study of the electrical activity.

Spontaneous electrical activity also occurs in insects. Read Roeder (1953, 1963) for additional information. Spontaneous activity in crayfish ganglia is discussed in Prosser (1934a) and Camougis and Kasprzak (1966). Electrical responses to light stimulation of the ganglia have been studied by Prosser (1934b) and Kennedy (1963). Welsh (1934) studied some of the behavioral aspects of photoreception by the tail gangion. Finally, neuropharmacological aspects of the crayfish ganglia, including actions on the light response, are discussed by Camougis (1967, 1968).

Chapter Six

Responses to Stimulation

Unlike the previous section in which we were concerned with the examination of a variety of spontaneous or endogenous action potentials, the experiments in this section will involve stimulation. The first two (Experiments 8 and 9) require stimulation of the receptors of an animal with the energy type that they normally detect in nature (light, mechanical touch, or contact). In the first instance the recording is of the bioelectrical transducer action of the photoreceptors themselves; in the second case, the responses are detected from the giant internuncial fibers in the central nervous system to which the sensory fibers transmit their signals through synapses. In the normal course of events, these giant fibers would connect with motor fibers going to the periphery so that appropriate motor responses would occur. These actions comprise the elements of the classical stimulus-response chain and they are shown diagrammatically in Fig. 16.

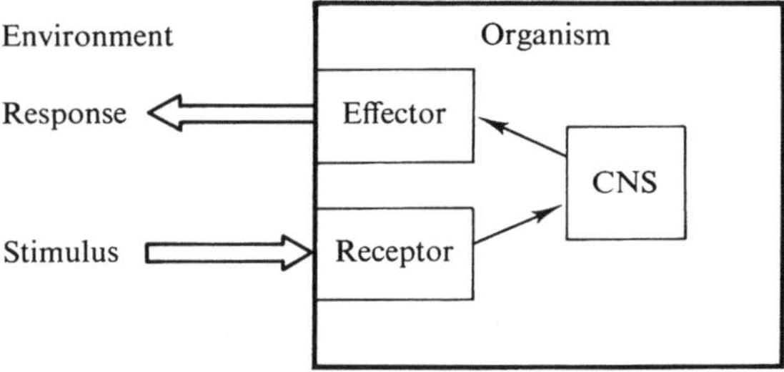

FIG. 16. Schematic diagram showing the elements of a stimulus-response chain in animals. In actuality the system is more complex; e.g., the CNS has centrifugal output back to the receptors.

In actual practice, of course, the system is much more complex, even for the lower animals. The interested reader is encouraged to read further in Granit (1955), Gray (1959), and Beament (1962). An early classic by Adrian (1928) is very readable and still very informative.

The final experiment is also the most complex in terms of instrumentation requirements because it involves electrical stimulation. Bioelectrical reponses to electrical stimulation are examined in the same cell. The isolated nerve is really a large number of long axons grouped together as a long biological cable. This preparation was studied extensively by Erlanger and Gasser, starting in the 1920's. Their work may be looked upon as the beginnings of modern electrophysiology. In Experiment 10 some of the classical aspects of the compound action potential may be studied. See the excellent monographs of Lorente de Nó (1947), Brazier (1960), Hodgkin (1964), Katz (1966), and Tasaki (1953, 1968) for further details. The reader interested in the historical aspects of the field will find hours of delightful reading in Adrian (1932), Hill (1932), Kato (1934), and Erlanger and Gasser (1937).

Photic Stimulation:
The Electroretinogram

The photoreceptors of both warm-blooded and cold-blooded animals are sensitive transducers in which incident photic energy will generate an electrical signal called the electroretinogram (ERG). The ERG is a somewhat complex, multiphasic wave which varies from species to species and with the experimental conditions. For our purposes, we will use a freshwater crayfish because of the convenience of preparation.

Using rubber bands, fasten to a small wooden board an adult crayfish about 3 to 5 inches long (dorsal side up). Insert a fine steel needle electrode along the dorso-lateral edge of the cornea of one eye. The axis of the needle electrode should be somewhat tangential rather than normal to the surface of the cornea at the point of entry, but it is essential that the needle actually pierce the corneal surface and not be merely in contact. This corneal electrode is the active lead. The indifferent electrode, which may be steel or platinum, can be placed at any convenient spot, preferably not on active tissue, for example, the dorsal surface of the cephalothorax. Drill a small hole on that surface and insert the indifferent electrode for a few millimeters, or enough to make good electrical contact with the underlying tissues. Do not use silver for either electrode, because photoelectric effects unrelated to the ERG may occur.

The electrophysiological apparatus will be similar to that described earlier (Fig. 14). The electrodes lead into a preamplifier which, in turn, leads into a CRO. The output of the preamplifier may also lead into a graphic recorder for permanent records of the ERG. The stimulus should be supplied by a d.c. light source. Since a rapid switching "on" and "off" can produce bothersome switching artifacts, it is best to turn on the light and to produce "on" and "off" effects on the eye by interrupting the beam mechanically. A simple flashlight held in a clamp will serve as a convenient stimulus source.

As the eye is illuminated, the corneal electrode will record a sharply rising negative wave with a peak amplitude of about 1 mV. The delay or latent period from the time the light goes "on" to the time the ERG starts to appear is about 25 msec. However, both the latent period and amplitude are dependent upon the stimulus intensity. Even with sustained illumination, the ERG starts to decay back to baseline after reaching its peak potential. Then, as the light goes "off", there is a smaller wave of positive potential and a final recovery to the original base level. The potential changes described make up the ERG of the crayfish. It is best to use very slow sweep frequencies (about 1/sec) at the beginning for detecting ERG responses.

The crayfish ERG is usually smooth and without "spiky" discharges. However, on occasion, a preparation will have a spiky activity, even when no illumination falls on the eye. The spiky discharges are usually muscle action potentials that are picked up by the recording electrodes. Such preparations are difficult to interpret, and, therefore, corrective actions should be taken. One remedy is to move the electrodes to less active regions, but this is not easy with the corneal electrode, since the main source of trouble with the corneal lead will probably be from movement of the eyestalk; usually, immobilizing it with non-toxic materials such as plaster of paris, while keeping the corneal surface free, will solve the problem.

After you are satisfied that you can record the ERG of the crayfish, you might try a series of additional experiments. One suggestion is to produce flickering light by interrupting the beam with a motor driven disc with sections cut out of its periphery. Then determine the highest frequency at which the ERG will follow before fusing: this is called the flicker fusion frequency (FFF). Another approach is to alter the light intensity by neutral density filters and then to determine such parameters as latency, peak amplitude, and the FFF. To carry out these suggestions you will need a good graphic recorder or oscilloscope camera because the measurements are difficult to make directly on the CRO display pattern without special procedures.

Camougis (1964) and Camougis and Kasprzak (1966) provide reference material on the crayfish ERG. Chapter 1 of Waterman (1961) and Chapter 5 of Granit (1955) and a special conference on photoreception (Wolken, 1958) will offer excellent references for those wishing to proceed further in this subject.

Mechanical Stimulation: Giant Fiber Discharges in Earthworms

In this experiment the electrical discharges of giant fibers in the CNS of the earthworm will be studied. These giant fibers are very large (Fig. 18) relative to other nerve fibers in the nerve cord, and, since the amplitude of the action potential under the recording conditions of this experiment varies directly with the fiber diameter, the giant fiber potentials are easily detected. It must be remembered that the giant fibers are not stimulated directly by mechanical contact with the body somites. The responses are mediated by synapses with sensory fibers located in the integument.

Anesthetize a large earthworm by placing it in 5 percent alcohol. After it is relatively still, pin the earthworm to a dissecting pan and carefully open it dorsally in the midbody region. Remove or retract the visceral mass to one side. Free the nerve cord from the surrounding tissue and carefully hook it with a recording electrode. Place the second (indifferent) electrode on any convenient, exposed tissue. Keep the preparation moist with frog Ringer solution. This is important—the preparation can dry out very quickly? If necessary, apply a little mineral oil on the exposed nerve. The mineral oil prevents drying but doesn't interfere with electrical recording. The electrodes are arranged to feed

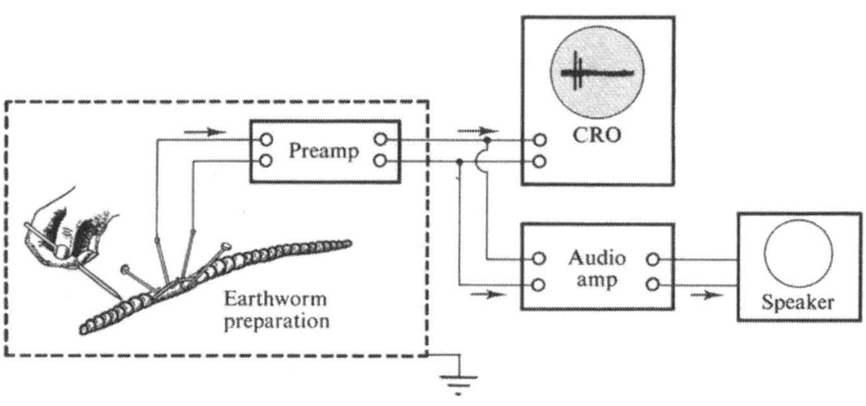

FIG. 17. Diagram of the experimental arrangement to study giant fiber discharges in the earthworm CNS following mechanical stimulation of the body surface.

into an amplifying and recording system. The sensitivity may be set at a few hundred μV/cm. The sweep of the CRO may be set at any convenient speed, preferably around 10 msec/cm.

The nerve cord may have a few spontaneously active units firing. Tap the anterior segments of the animal with a fine glass or wooden rod (Fig. 17). This should generate a number of action potentials in the giant fibers of the cord (Fig. 18). Note that the amplitude of these spikes is far greater than that of any spontaneous activity. Try stimulating the rear segments as well. Determine (1) whether there is a difference in the recorded activity when stimulating the anterior or posterior end, and (2) what are the sensory fields (number of segments) at the two ends which produce spikes.

Receptors are very sensitive to peripheral blocking agents. Inject a small amount of lidocaine or procaine (1 or 2 percent) under the epidermis at one end. Again, tap the segments at the injection site and note any differences. In a similar manner, study the block of conduction in the cord itself by placing the blocking agent between stimulus and recording sites. Determine if the blocking action is reversible. Also, determine what other sensory modalities (chemical or physical) will generate action potentials in the giant nerve fibers. For an excellent account of giant fiber systems in annelids, the reader is urged to read pages 689–708 of Bullock and Horridge (1965).

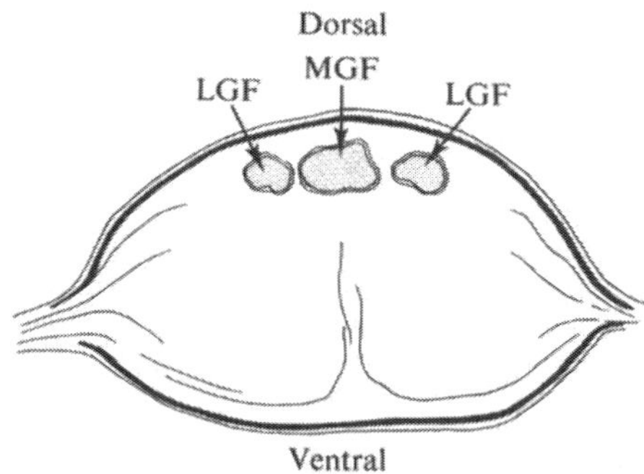

FIG. 18. Cross-section of the ventral nerve cord (CNS) of the earthworm showing the medial giant fiber (MGF) and two lateral giant fibers (LGF) in the dorsal region.

Electrical Stimulation:
Compound Action Potentials in Peripheral Nerve

The study of the compound action potential from isolated frog nerve with electrophysiological methods is the classical approach to the study of neurophysiology. This is the experiment that one sees described in textbooks, and is, perhaps, the most familiar to students. In concept the procedure is simple, consisting of stimulation at one end of the nerve and recording the propagated action potential at the other end. This is similar to the previous experiment, the exception being that electrical rather than physiological stimulation is applied to the preparation. Consequently, additional electrodes and apparatus are involved. The instrumentation system is illustrated in Fig. 6.

The sciatic-peroneal nerve of the frog is used in this experiment. Pith a frog and remove the skin from the abdomen and hind limbs by peeling it over itself in the posterior direction. The carcass is then pinned on a dissection board, dorsal side up. Carefully expose the spinal nerves emerging posteriorly from the pelvic region. Separate the muscles of the thigh region and follow the prominent sciatic nerve to the knee. Various branchings of the nerve will be observed. Follow the largest trunk below the knee to about the end of the tibiofibula bone: this will be the sciatic-peroneal nerve. While the nerve is still exposed but in place, tie a thread to each end, making the nerve preparation as long as possible. Next, cut the nerve close to the thread proximally and distally and at the branches as appropriate. By means of the threads, the nerve preparation can be manipulated with minimal handling of the nerve itself. After cutting and freeing the nerve, place it immediately in a bath of frog Ringer solution. From an average sized frog it should be possible to get a nerve preparation at least 7 to 10 cm in length. If excess fascia or other tissue remain on the nerve, clean them off by securing the nerve with one thread and using fine forceps to remove the unwanted tissue. The nerve is then ready for mounting on the electrodes.

Two contradictory requirements for the experiment must be considered in the design of the nerve chamber and its electrodes. First, the nerve must be kept constantly moistened with Ringer solution so that it remains viable for many hours, and second, it must be free of any fluid at the actual time of stimulation and recording; otherwise, both the stimulus and action potentials will be shunted out. Thus, the nerve chamber must permit bathing of the nerve and a means of removing the solution from the nerve when stimulating and recording.

57

Also, an enclosed chamber will offer advantages in minimizing evaporation of moisture from the nerve while the nerve is actually out of solution. A simple chamber is shown in Fig. 4.

Place the nerve on the series of electrodes spaced about 1 cm apart and secure it in place with the threads. Avoid excessive tension on the nerve; enough tension to break the surface tension of the bathing fluid is satisfactory. The stimulating electrodes should be at the thick (proximal) end of the nerve. A ground electrode is located near the middle of the nerve, while the recording electrodes are at the distal end of the nerve. Connect the stimulating electrodes to the output side of the stimulator (or stimulus isolation unit if required) in such a way that the cathode is nearest the recording electrodes. Connect a middle electrode to ground, and, finally, connect two distal electrodes to the input side of the preamplifier. Now the experimentation may proceed.

Notice that the stimulator allows the investigator to change three stimulus parameters: voltage, pulse duration, and frequency. As a beginning, set the voltage at about 1 volt (this may vary considerably), duration at 0.1 msec, and the frequency at 30 pulses/sec. The sensitivity of the amplification and display system should be set at about 1 mV/cm. The sweep speed may vary; however, 1 msec/cm will provide a good display of the action potential. The delay between the triggering signal to the CRO and the stimulus may be set at about 2 msec. Other settings will depend upon the specific instruments used. In general, they are set to values consistent with good recording display and not adjusted again during the experiment.

Turn on the stimulator; stimulation should produce a display on the screen of the CRO. If the display is not recognized immediately as the compound action potential, vary the voltage of the stimulator up and down, which should produce small changes in the stimulus artifact (Fig. 21) and large changes in the compound action potential. If both recording electrodes are on healthy nerve, a biphasic action potential will appear. It is easier to conduct experiments with a monophasic action potential, which is accomplished by pinching the nerve between two recording electrodes. After the monophasic potential is produced, reduce the voltage until the first sign of an action potential appears: this is the *threshold* value for the most sensitive nerve fibers. Since the nerve preparation contains thousands of individual fibers, its threshold value cannot be described with full accuracy. Hence the threshold is defined as the voltage value required to achieve stimulation of just sufficient units to be detected under the conditions of the experiment.

After the action potential is displayed and the threshold determined, increase the voltage until the amplitude ceases to increase. Voltage values above this point are called *supramaximal* stimuli. Study the amplitude of the action potential as a function of stimulus voltage. Convince yourself that the action potential has a definite *polarity* relative to inactive nerve by interchanging the connections of the two recording electrodes with the preamplifier.

The voltage required to stimulate the nerve depends upon the duration of the stimulus pulse. This can be seen readily by varying the pulse duration and determining the new threshold each time. The same criterion for a threshold response must be observed carefully. Table 4 and Fig. 19 show the data for such an actual experiment with a frog sciatic-peroneal nerve preparation. The graph shown in Fig. 19 is known as a *strength-duration curve*. Determine such a strength-duration curve for your preparation.

TABLE 4. ACTUAL STRENGTH-DURATION DATA FROM AN ISO-LATED FROG SCIATIC-PERONEAL NERVE PREPARATION

Strength in V	Duration in msec
0.52	0.02
0.40	0.04
0.32	0.06
0.27	0.08
0.23	0.1
0.10	0.2
0.075	0.4
0.065	0.6
0.057	0.8

Let us now vary stimulus frequency, the third stimulus parameter. Set the pulse duration at 0.1 msec and the voltage at a supramaximal strength. Slowly increase the stimulus frequency above 30 pulses/sec. The action potential should appear relatively the same up to a frequency of at least 100 pulses/sec. Continue to increase the frequency. Note that the amplitude of the action potential decreases as the frequency approaches 200–300 pulses/sec. This is often referred to as *high frequency failure*. The reason for this phenomenon is that some fibers are not able to recover in time for the next impulse since they are in their *refractory period* when the next stimulus pulse occurs. The higher the frequency the greater the number of fibers which cannot follow the stimulus frequency. Hence the amplitude of the compound action potential decreases, since it is the summation of many individual action potentials. Increasing the voltage beyond the supramaximal value will not increase the amplitude of a nerve undergoing high frequency failure. Find the frequency at which high frequency failure occurs in your nerve preparation.

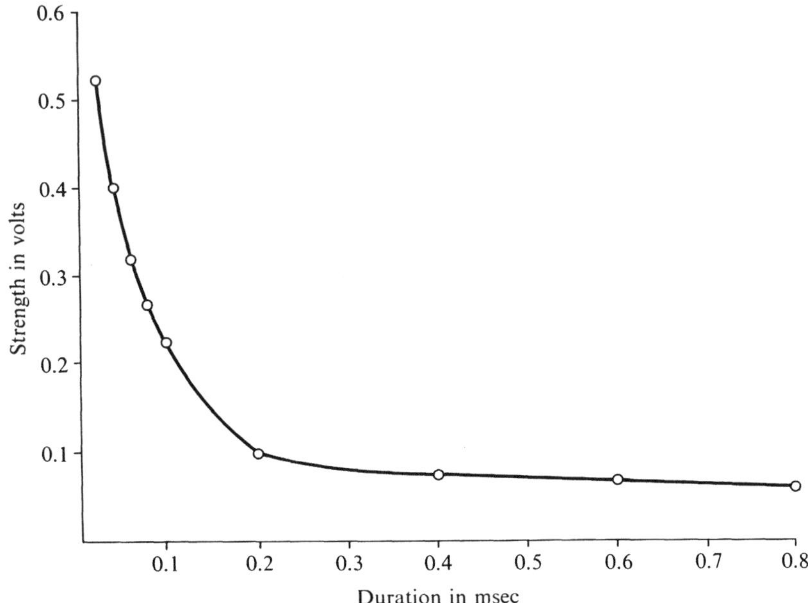

FIG. 19. Strength-duration curve of actual data from an isolated sciatic-peroneal nerve preparation from a frog.

The propagated action potential has a definite *conduction velocity* which can be measured without great difficulty. The formula for average velocity is

$$v = \frac{d}{t}$$

where v = average velocity, d = distance, and t = time.

In our case distance is the value for the nerve length between the cathode stimulus electrode and the first recording electrode. The time can be measured on the scope as the distance between the stimulus artifact and the peak deflection of the compound action potential. Since sweep speed should be known, distance is translated into time. Then the average velocity may be calculated. In neurophysiology, conduction velocities by convention are given in m/sec. In good preparations velocities of 10 to 30 m/sec are measured routinely. Measure the average conduction velocity of your nerve preparation. Is the velocity uniform throughout the length of the nerve?

A final phenomenon to study is that of conduction block by local anesthetics, compounds which permit reversible block of action potentials in excitable tissues. Devise a small drug chamber about 1 cm across from plastic or wax. This chamber should fit between the stimulating and recording electrodes in such a way that the nerve can pass over it. After a good monophasic potential is displayed on the CRO, fill the drug chamber with 0.5 percent lidocaine or procaine in frog Ringer solution. The nerve must be in good contact with the meniscus of the drug solution. Note the rapid action of the drug as evidenced by a decline of the amplitude of the action potential. After the amplitude has declined to about 50 per cent of its control value, quickly wash the nerve in frog Ringer solution. Note the recovery of the nerve. Plot the amplitude versus time; this shows the block and recovery of the nerve in graphic form. Repeat this procedure several times to show that the actions of local anesthetic drugs are reversible.

Consult a standard textbook for a discussion on nerve action potentials. Brazier (1960) and Florey (1966) also have excellent accounts of action potentials in peripheral nerve. Hodgkin (1964), Katz (1966), and Tasaki (1968) give somewhat more advanced discussions of the mechanisms underlying the nerve impulse. Camougis and Takman (in press) discuss methods for testing local anesthetics on nerve tissue.

Chapter Seven

Notes on the Analysis and Interpretation of Electrical Signals

One of the biggest problems encountered by investigators who are beginners in electrophysiology is the interpretation of results. Often the first aspect of the problem centers on whether the output of whatever read-out device is used is, indeed, related to biological phenomena. Therefore, one of the first considerations is to determine whether the output signals are biological phenomena related to the experiment or some sort of artifact. This sounds fundamental, but it cannot be overstressed. One thing that will be extremely helpful if it is constantly in mind is the analysis of the axes of the read-out device. The approach of electrophysiology is fundamentally to make an electrographic analysis or recording. Consequently, whether one uses a CRO, a camera, digital voltmeter, or tape recorder, one axis (usually the vertical) is always a voltage axis. In addition, since most signals vary as a function of time, the other axis is a time axis. Again, most instruments are set up such that the horizontal axis is a time axis, but a digital voltmeter, of course, would be an exception. Since we have the two axes (voltage and time) as an output, the first consideration is to examine the signal critically with respect to these two parameters. Does the output make sense? Even the novice investigator has a good idea of the amplitude and temporal dimensions of the signals he expects to find. Therefore, a critical examination of the voltage and time axes will help determine whether the signal is reasonable for the experimental conditions. It is not by accident that editors demand clearly labelled axes on electrographic records published in their journals.

Should the signals be unreasonable or even obviously artifacts, the next aspect of the problem is to determine the source. As with anything involving technique, experience plays a major role in solving technical problems. However, as a guide, the more common sources of artifacts or noise have been listed in Table 5. It must be remembered that this is only a partial list, and that many bizarre signals may be picked up, often baffling even experienced investigators.

TABLE 5. A LIST OF COMMON ARTIFACTS AND THEIR POSSIBLE SOURCES ENCOUNTERED DURING ELECTROPHYSIOLOGICAL EXPERIMENTATION

Appearance of artifact	Possible source
60 Hz sinusoidal waves	Pick up from a.c. power mains or extensions
High frequency signals (> 1000 Hz)	Fluorescent lamps Radio or TV transmissions Electronic equipment
High frequency static	Poor connection
Low frequency waves (< 10 Hz) appearing as big shifts in baseline	Differential hydration or evaporation at recording electrodes
Slow d.c. shifts	Polarization of electrodes

Once the source of the artifact problem is located, the next step, obviously, is its elimination. Here again, considerable experience may be required, but certain rules might be followed to good advantage. The most common type of interference is from a 60 Hz a.c. pickup from power mains, and its most probable site is on the input side of the preamplifier. Therefore, the investigator should check on the following points:

(1) adequacy of the shielded cage

(2) proper grounding

(3) good contact between electrodes and tissue

(4) all electrical contacts between electrodes, input leads, and instruments (preamplifier especially).

If good shielding, grounding and connections are part of the technique, very few problems with 60 Hz a.c. artifacts will be encountered. Figure 20 illustrates some of the points just discussed.

The other problems can be corrected in a number of ways, but shielding and grounding are also important for eliminating high frequency pickup. Sometimes other laboratory equipment, such as oscillators or incubators, may produce the source of high frequency pickup; such interference can be eliminated by changing the location of the recording station or simply by turning off the interfering instruments. Very low frequency signals can be corrected by keeping the

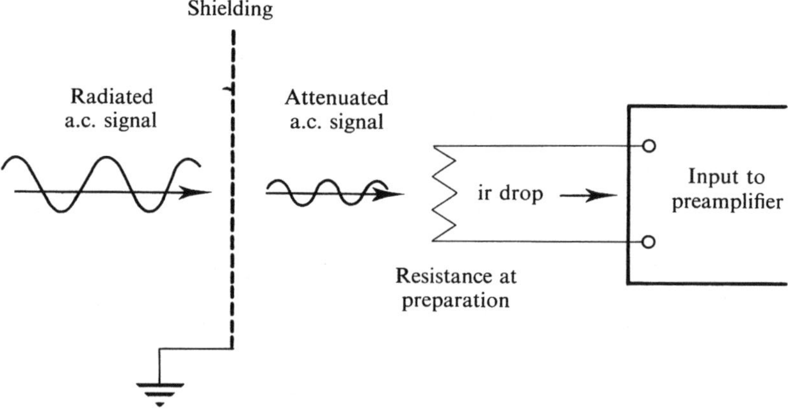

FIG. 20. Shematic diagram of the attenuation of an a.c. artifact by shielding and grounding. The resulting signal causes some interference at the preparation, but it is small compared with the bioelectrical signal. Also, it is further reduced through differential input to the preamplifier.

preparation in a chamber where the moisture quickly comes to equilibrium with the preparation. Polarization effects can be reduced by using non-polarizing electrodes. The alert reader will conclude at this point that the trouble shooting is empirical and mainly common sense: this is, indeed, the case. In concluding the discussion on artifacts, we can state the following principles: check the entire instrumentation system; use common sense; don't get discouraged. Practice and persistence are necessary for learning any new technique. For those intending to pursue electrophysiological techniques in their research work, it is recommended that they read more extensively on the subject of noise and interference. Appropriate sections in Donaldson (1958), Welsh and Smith (1960), Suckling (1961), Kay (1964), Nastuk (1964), and Yanof (1965) will serve as a good beginning.

Let us now turn to the bioelectric signals, representing the biological phenomena. Again, the analysis requires an examination of the two axes of the electrographic recordings or outputs. Such an examination can tell us a lot about our signals, as for example:

(1) amplitude
(2) duration
(3) waveform
(4) periodicity or apparent randomness
(5) frequency
(6) pulse repetition rate.

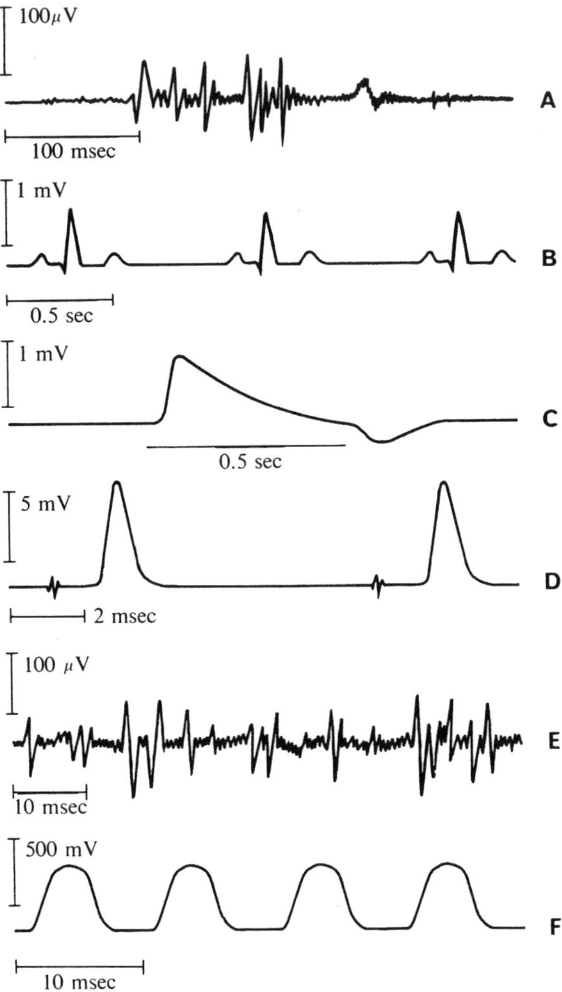

Fig. 21. Diagrams of a variety of electrographic recordings of bioelectrical potentials.

A. EMG (non-specific). B. EKG from a mammal. C. ERG from a crayfish; time line of stimulus shows duration. D. Compound action potential from peripheral nerve (e.g., frog or rat); note stimulus artifact preceding the compound action potential. E. Spontaneous action potential discharges from arthropod ganglia. F. Electric organ discharges from a weakly electric fish.

Voltage and time axes are labelled in each case.

Therefore, certain quantitative measurements about the bioelectrical signals are readily possible. As an example, with Experiment 10, the experimenter might determine that the amplitude of the A spike is 7.5 mV, the duration is 1.5 msec, and the nerve can be driven at about 150 pulses/sec before there is a loss in amplitude of the signal. Alternatively, the investigator doing Experiment 5 might conclude that the electromyographic discharges from the human hand appear random. The amplitude of the signals is several hundred μV for the large spikes and the frequency varies considerably. Finally, the person doing Experiment 4 may be struck by the regularity of the sinusoidal waveform coming from the electric fish from which he is recording. All these conclusions attest to the fascinating nature and variety of bioelectrical signals.

Before concluding this section we should consider the positions of the electrodes relative to each other and to the preparation. The polarity of the electrodes should always be known. If stimulating electrodes are used, the positions of the cathode and the anode should be known. Also, it should be remembered that the cathode is responsible for the simulation. Therefore, it is possible to make an inference as to where the electrically induced signals originated. Similarly, the recording electrodes are connected ultimately to the input side of a preamplifier or meter of some type. The investigator should know which lead is active and which goes to ground, or alternatively, which is connected to a positive and which to a negative input terminal. Then it is easy to state that when a certain electrode is negative with respect to another, the recording system shows an upward deflection. This, in turn, permits an analysis of the experimental results in terms of relative positivity and negativity at the electrodes as a function of time. Thus, both tonic and phasic signals can be studied in relation to structure and electrode locations. Figure 21 illustrates a variety of electrographic recordings of bioelectrical action potentials.

Part Four

REFERENCES

1. Literature Cited

Adrian, E. D. 1928. *The Basis of Sensation*. London: Christophers. (Reprinted and Published in 1964 by Hafner Publishing Co., Inc., New York.)

————. 1932. *The Mechanism of Nervous Action*. Philadelphia: University of Pennsylvania Press.

American Physiological Society. 1967. *Laboratory Experiments in General Physiology*. Rev. ed. Washington, D.C.

Bayley, R. H. 1958. *Electrocardiographic Analysis*, vol. 1, *Biophysical Principles of Electrocardiography*. New York: Hoeber.

Beament, J. W. L., ed. 1962. "Biological Receptor Mechanisms," Number XVI, *Symposia of the Society for Experimental Biology*. New York: Academic Press.

Becker, R. O. 1960. "The bioelectric field pattern in the salamander and its simulation by an electronic analog." *IRE Trans. Med. Electronics*, ME-7(3): 202–207.

Brazier, M. A. B. 1959. "The historical development of neurophysiology." In Field, J., ed.-in-chief, *Handbook of Physiology*. Magoun, H. W., section ed., Section 1, *Neurophysiology*. Vol. 1, 1–58. Washington, D.C.: American Physiological Society.

————. 1960. *The Electrical Activity of the Nervous System*. 2nd ed. New York: Macmillan.

Brown, C. C., and Saucer, R. T. 1958. *Electronic Instrumentation for the Behavioral Sciences*. Springfield, Illinois: C. C. Thomas.

Brown, C. C., and Webb, G. N. 1964. *Instrumentation with Semiconductors for Medical Researchers*. Springfield, Illinois: C. C. Thomas.

Bullock, T. H., and Horridge, G. A. 1965. *Structure and Functions in the Nervous Systems of Invertebrates*. Vol. 1. San Francisco: Freeman.

Bures, J., Petran, M., and Zachar, J. 1967. *Electrophysiological Methods in Biological Research*. 3rd ed. New York: Academic Press.

Burns, B. D., ed. 1961. Section IV. "Electrical recording from the nervous system." In Quastel, J. H., ed.-in-chief. *Methods in Medical Research*. Vol. 9. Chicago: Year Book Medical Publishers.

Camishion, R. C. 1964. *Basic Medical Electronics*. Boston: Little, Brown.

Camougis, G. 1960a. "Recording bioelectrical potentials from aquatic animals." *Turtox News*. 38: 156–157.

———. 1960b. "Visual responses in crayfish. I. Recording shock responses to light with remote electrodes." *J. Cell. Comp. Physiol*. 55: 189–194.

———. 1964. "Visual responses in crayfish. II. Central transmission and integration." *J. Cell. Comp. Physiol*. 63: 339–352.

———. 1967. "Unmasking of a photoinhibitory response in the crayfish caudal ganglion following partial anesthesia." *Comp. Biochem. Physiol*. 21: 231–236.

———. 1968. "Neuropharmacology of invertebrates: Quantitative studies with an isolated CNS preparation." In Efron, D. H., ed.-in-chief. *Psychopharmacology, A Review of Progress 1957–1967*. Public Health Service Publication, No. 1836. Washington D.C.: U.S. Government Printing Office.

Camougis, G., and Kasprzak, H. 1966. "Visual responses in crayfish. III. Further studies on transmission through the brain." *J. Cell Physiol*. 67: 45–52.

Camougis, G., and Takman, B. H. "Methods for evaluating local anesthetics." In Schwartz, A., ed. *Methods in Pharmacology*. New York: Appleton-Century-Crofts (in press).

Catton, W. T. 1957. *Physical Methods in Physiology*. London: Pitman.

Chagas, C., and Paes de Carvalho, A., eds. 1961. *Bioelectrogenesis*. New York: Elsevier.

Dickinson, C. J. 1950. *Electrophysiological Technique*. London: Electronic Engineering.

Donaldson, P. E. K. 1958. *Electronic Apparatus for Biological Research*. New York: Academic Press.

Eccles, J. C. 1964. *The Physiology of Synapses*. Berlin: Springer-Verlag.

Erlanger, J., and Gasser, H. S. 1937. *Electrical Signs of Nervous Activity*. Philadelphia: University of Pennsylvania Press.

Florey, E. 1966. *An Introduction to General and Comparative Animal Physiology*. Philadelphia: Saunders.

Fogel, L. J. 1963. *Biotechnology: Concepts and Applications*. Englewood Cliffs, New Jersey: Prentice-Hall.

Galambos, R. 1962. *Nerves and Muscles: An Introduction to Biophysics*. Garden City, New York: Anchor Books, Doubleday.

Geddes, L. A., and Baker, L. E. 1968. *Principles of Applied Biomedical Instrumentation*. New York: Wiley.

Goldstein, N. N., Jr. 1964. *Instrumentation Methods for Physiological Studies*. Vol. 1. Regents of the University of California (Reprinted by Heath Company, Benton Harbor, Michigan).

Granath, L. P., Erskine, F. T., III, Maccabee, B. S., and Sachs, H. G. 1968. "Electric field measurements on a weakly electric fish." *Biophysik.* 4: 370–372.

Granath, L. P., Sachs, H. G., and Erskine, F. T., III. 1967. "Electrical sensitivity of a weakly electric fish." *Life Sciences.* 6: 2372–2377.

Granit, R. 1955. *Receptors and Sensory Perception.* New Haven: Yale University Press.

Gray, J. A. B. 1959. "Initiation of impulses at receptors." In Field, J., ed.-in-chief, *Handbook of Physiology*, Magoun, H. W., section ed., Section 1, *Neurophysiology*. Vol. 1, 123–145. Washington, D.C.: American Physiological Society.

Green, R. M. 1953. "Commentary on the Effect of Electricity on Muscular Motion." A translation of Luigi Galvani's *De Viribus Electricitatis in Motu Musculari Commentarius.* Cambridge: Elizabeth Licht.

Grundfest, H. 1960 October. "Electric fishes." *Scientific American.* 203: 115–124.

Hainsworth, F. R., Camougis, G., and Granath, L. P. 1963. "Threshold values for electroreception in some gymnotids." *Am. Zool.* 3: 483.

Hecht, H. H., ed. 1957. The electrophysiology of the heart. *Ann. N.Y. Acad. Sci.* 65, Art. 6: 653–1146.

Hill, A. V. 1932. *Chemical Wave Transmission in Nerve.* New York: Macmillan.

Hill, D. W. 1965. *Principles of Electronics in Medical Research.* Washington, D.C.: Butterworth's.

Hodgkin, A. L. 1964. *The Conduction of the Nerve Impulse.* Liverpool: Liverpool University Press.

Hoff, H. E., and Geddes, L. A. 1967. *Experimental Physiology.* 3rd ed. Houston: Baylor University College of Medicine (Distributed by E and M Instrument Co., Inc., Houston, Texas).

Hoffman, B. F., and Cranefield, P. F. 1960. *Electrophysiology of the Heart.* New York: McGraw-Hill.

Kato, G. 1934. *The Microphysiology of Nerve.* Tokyo: Maruzen.

Katz, B. 1966. *Nerve, Muscle, and Synapse.* New York: McGraw-Hill.

Kay, R. H. 1964. *Experimental Biology: Measurement and Analysis.* New York: Reinhold.

Kennedy, D. 1963. "Physiology of photoreceptor neurons in the abdominal nerve cord of the crayfish." *J. Gen. Physiol.* 46: 551–572.

Lion, K. S. 1957. *Instrumentation in Scientific Research: Electrical Input Transducers.* New York: McGraw-Hill.

Lissmann, H. W. 1958. "On the function and evolution of electric organs in fish." *J. Exp. Biol.* 35: 156–191.

———. 1963 March. "Electric location by fishes." *Scientific American.* 208: 50–59.

Lorente de Nó, R. 1947. *A Study of Nerve Physiology.* New York: Rockefeller Institute.

McLennan, H. 1963. *Synaptic Transmission.* Philadelphia: Saunders.

Nastuk, W. L., ed. 1963. *Physical Techniques in Biological Research.* Vol. 6. *Electrophysiological Methods*, Part B. New York: Academic Press.

———, ed. 1964. *Physical Techniques in Biological Research.* Vol. 5. *Electrophysiological Methods*, Part A. New York: Academic Press.

Newman, D. W., ed. 1964. *Instrumental Methods of Experimental Biology.* New York: Macmillan.

Ochs, S. 1965. *Elements of Neurophysiology.* New York: Wiley.

Offner, F. 1967. *Electronics for Biologists.* New York: McGraw-Hill.

Phillips, L. F. 1966. *Electronics for Experimenters in Chemistry, Physics, and Biology.* New York: Wiley.

Prosser, C. L. 1934a. "Action potentials in the nervous system of the crayfish. I. Spontaneous impulses." *J. Cell. Comp. Physiol.* 4: 185–209.

———. 1934b. "Action potentials in the nervous system of the crayfish. II. Responses to illumination of the eye and caudal ganglion." *J. Cell. Comp. Physiol.* 4: 363–377.

Roeder, K. D., ed. 1953. *Insect Physiology.* New York: Wiley.

———. 1963. *Nerve Cells and Insect Behavior.* Cambridge: Harvard University Press.

Rosenblith, W. A., ed. 1962. *Processing Neuroelectric Data.* Cambridge: M.I.T. Press.

Ruch, T. C., Patton, H. D., Woodbury, J. W., and Towe, A. L. 1961. *Neurophysiology.* Philadelphia: Saunders.

Rushmer, R. F., ed.-in-chief. 1966. *Methods in Medical Research.* Vol. 11. Chicago: Year Book Medical Publishers.

Smith, O. A., Jr., ed. 1966. Section V, *Neurophysiologic Technics.* In Rushmer, R. F., ed.-in-chief, *Methods in Medical Research.* Vol. 11, Chicago: Year Book Medical Publishers.

Stacy, R. W. 1960. *Biological and Medical Electronics.* New York: McGraw-Hill.

Stevens, C. F. 1966. *Neurophysiology: A Primer.* New York: Wiley.

Suckling, E. E. 1961. *Bioelectricity.* New York: McGraw-Hill.

Suckling, E. E., and Koizumi, K. 1967. "An Electrophysiological Teaching Laboratory." *Trans. N.Y. Acad. Sci.* Ser. II. Vol. 29. No. 7: 903–910.

Suprynowicz, V. A. 1966. *Introduction to Electronics: For Students of Biology, Chemistry, and Medicine.* Reading, Massachusetts: Addison-Wesley.

Tasaki, I. 1953. *Nervous Transmission.* Springfield, Illinois: C. C. Thomas.

———. 1968. *Nerve Excitation: A Macromolecular Approach.* Springfield, Illinois: C. C. Thomas.

Tolles, W. E., ed. 1964a. "Electronics in the medical specialties." *Ann. N.Y. Acad. Sci.* 118. Art. 1: 1–133.

———, ed. 1964b. "Computers in medicine and biology." *Ann. N.Y. Acad. Sci.* 115. Art. 2: 543–1140.

Waterman, T. H., ed. 1961. *The Physiology of Crustacea.* Vol. 2. New York: Academic Press.

Welsh, J. H. 1934. "The caudal photoreceptor and responses in the crayfish to light." *J. Cell. Comp. Physiol.* 4: 379–388.

Welsh, J. H., and Smith, R. I. 1960. *Laboratory Exercises in Invertebrate Physiology.* rev. ed., Appendix V., *Electrophysiological Methods.* Minneapolis: Burgess.

Whitfield, I. C. 1953. *An Introduction to Electronics for Physiological Workers.* New York: Macmillan.

———. 1964. *Manual of Experimental Electrophysiology.* New York: Macmillan.

Wolken, J. J., ed. 1958. "Photoreception." *Ann. N.Y. Acad. Sci.,* 74. Art. 2: 161–406.

Wooldridge, D. E. 1963. *Machinery of the Brain.* New York: McGraw-Hill.

Yanof, H. M. 1965. *Biomedical Electronics.* Philadelphia: F. A. Davis.

2. *Additional Reference Material*

In addition to the references given above, useful reference material may come from a variety of sources. These will be listed below in a series of categories.

(1) Buyers' Guides

Various buyers' guides are published periodically, usually on an annual basis. These offer convenient surveys of available equipment.

Electronics Buyers' Guide. (McGraw-Hill.)

Instrument and Control Systems, Buyers' Guide Issue. (Rimbach Publications.)

Laboratory Guide to Instruments, Equipment, and Chemicals. (American Chemical Society.)

Medical Instrument Dictionary and Buyers' Guide. (Rimbach Publications.)

Science Guide to Scientific Instruments. (American Association for the Advancement of Science.)

(2) Instrument Surveys

Publishers in the instrument field make periodic surveys of different classes of instruments. Examples appear below.

Instrumentation Review, in: *Electrical Equipment*, February 1967. (Sutton Publishing Company.)

Oscilloscope Survey, in: *Instrument and Control Systems*, January 1968. (Rimbach Publications.)

Recorder Manual, Aronson, M. H. (editor), 1965. A series of papers including a survey on graphic recorders. (Rimbach Publications.)

(3) Pertinent Literature from Industrial Companies

Some of the instrument manufacturers publish and circulate material that is useful to electrophysiological investigators and others interested in biomedical instrumentation.

Argonaut Associates, Inc., *Insight*. (A forum and idea exchange about medical electronics instruments.)

Beckman Instruments, Inc., *News of Physiological Instrumentation*.

Leeds and Northrup, *Technical Journal*. (Mainly for engineers but some articles of interest to biologists.)

Sanborn Division, Hewlett-Packard Company, *Measuring for Medicine and the Life Sciences.*
Tektronix, Inc., *Notes on the Practical Photography of Oscilloscope Displays. Fundamentals of Selecting and Using Oscilloscopes. TEKScope;* formerly *Service Scope.* (Published periodically; some articles of interest to biomedical instrumentation).

(4) Periodicals

In recent years a number of periodicals specializing in biomedical engineering and instrumentation have been published. Many of these publish frequent articles in electrophysiology. Examples of such periodicals include the following:

Bio-Medical Engineering,

IEEE Transactions on Bio-Medical Engineering,

Medical and Biological Engineering,

Medical Research Engineering.

Also, it should be stressed that numerous, excellent papers are published yearly, both in techniques and theory, in the standard journals sponsored by the professional societies. Investigators progressing beyond the earliest stages should consult such journals as a matter of routine.

In concluding this section on literature, the author wishes to stress that the reference sources given do not necessarily make up an exhaustive survey. The scientific fields of electrophysiology and pertinent instrumentation are very active, and the literature available is voluminous. Any omissions of important reviews, books, surveys, or periodicals do not reflect on any authors, manufacturers, or publishers. They merely represent gaps in the experience and knowledge of the author.

Part Five

APPENDICES

Physiological Solutions

FROG RINGER SOLUTION

Solute	g/liter	mM/liter
NaCl	6.5	111.2
KCl	0.14	1.88
$CaCl_2$	0.12	1.08
$NaHCO_3$	0.2	2.38
NaH_2PO_4	0.01	0.08
Glucose	2.0	11.11

CRAYFISH SOLUTION (VAN HARREVELD'S SOLUTION)

Solute	g/liter	mM/liter
NaCl	12.0	205.3
KCl	0.4	5.37
$CaCl_2$	1.5	13.55
$MgCl_2.6H_2O$	0.53	2.61
$NaHCO_3$	0.2	2.38

The $NaHCO_3$ should be dissolved before adding the other compounds; this will prevent precipitation.

Hints on the Selection of Electrophysiological Apparatus

There are no "rules" governing the selection of scientific apparatus. Very often personal and subjective matters enter into selection as much as the realities of the scientific applications. Also, the scientific applications outnumber the available models, indicating that versatility must be inherent in some instruments. Versatility implies potential application to several problems, and this, in turn, implies a potential economy. These are all important considerations. In this appendix certain considerations based on both personal experience and that of various colleagues are listed with the hope that they will be useful hints on the selection of equipment.

Application. Know what you wish to do with a particular instrument. This point was also stressed in Part Two above. First determine whether the instrument is to be used for teaching, a combination of teaching and research projects for students, or research by senior people. Next define the specific application. Remember that a complex CRO, for example, is not ideal for teaching inexperienced students. On the other hand, instruments used for general teaching one semester may not be sophisticated enough for some student research projects during the next semester. However, if one had to define the direction in which the selection process goes, one would have to tend towards the use of instruments more complex than warranted by the experiments being done. This is not always the fault of the person making the selection: the alternatives may all include instruments that are over-engineered for that particular application.

Information. The next step is to get information on instruments available for your application. Buyer's guides, manufactureres' catalogs, surveys, sales engineers, and colleagues are good sources of information. Never hesitate to ask plenty of questions: enthusiastic replies will result, especially from those who are quite expert in this area. Annual meetings of professional societies offer excellent opportunities for contact with suppliers, literature, and people who can help supply information, of which the more at your disposal, the more sound will be your final decision.

Specifications. Once the information is available, study the specifications in the light of your application requirements. The specifications discussed in Part Two are minimal considerations for most electrophysiological techniques. If any of the specifications are unclear, have a sales engineer explain them to you. It is his job. Avoid paying for special features that you don't need for your specific applications.

Future Applications. Try to anticipate future needs. Obviously, precise future needs cannot be determined, but some predictions are possible. For example, when buying a graphic recorder, it might be useful to obtain a chassis capable of holding four channels, even though only two are installed initially. This allows for the expansion of a research problem. As another example, if work with intracellular electrodes is anticipated, the read-out devices should be d.c.-coupled. Finally, it should be borne in mind that if an academic department is starting an independent studies program, almost certainly the instrument requirements will change.

Cost. Brief mention of this practical problem is necessary. Table 2 gives a good indication of how costs may vary. Two decades of budgetary affluence have resulted in overspending in many laboratories. Government, college, and industrial laboratories have all been guilty of lavish outlays for scientific equipment. For teaching undergraduate and secondary school courses, the problem may be more one of restricted funds. Whatever the fiscal situation, considerations of cost may modify the final decision. Remember, too, that an electrophysiology laboratory will also have need for other equipment such as balances, pH meters, test equipment, etc. Payment for such additional apparatus could come from realistic reductions in outlay after considering the needs, specifications, and costs of electrophysiological equipment.

Maintenance. Maintenance, calibration, and repair of equipment poses another practical consideration. Large research groups may have a technician or engineer who is an expert in electronics and who can do all routine maintenance. However, a school wishing to do a few exercises in electrophysiology once a year has other problems. An electronics workshop is justified only on a volume basis. Commercial firms specializing in equipment maintenance are available in most larger metropolitan areas; however, their services are expensive. There is no single solution to this problem, but it should not be forgotten when buying equipment.

Storage. Storage poses another practical consideration. Again, most research groups probably have their equipment at recording stations year round.

Other organizations, including some schools and some industrial companies, may dismantle the instrumentation system after a semester or completion of a project. Large multi-channel recorders and oscilloscopes are bulky, and if they are not properly stored they may receive abuse and even suffer missing parts. Consider this important question before ordering your equipment.

Time Factors. A final point to consider is the time required for various phases centering around the acquisition of equipment. It takes time to get the information required. Some apparatus may require at least 6–12 months delivery time after an order is placed. Such simple things as getting platinum and silver wire may take a long time. The making up of nerve chambers at local shops may take some considerable time. Start planning and studying your equipment needs about a year in advance of when it is hoped to commence the experiments. This investment in time will bring many returns when it comes to engage in the electrophysiology experiments, not the least of which will be a better understanding of your instrumentation system.